AF806125

Lecture Notes in Control and Information Sciences

Volume 472

Series editors

Frank Allgöwer, Stuttgart, Germany
Manfred Morari, Zürich, Switzerland

Series Advisory Boards

P. Fleming, University of Sheffield, UK
P. Kokotovic, University of California, Santa Barbara, CA, USA
A.B. Kurzhanski, Moscow State University, Russia
H. Kwakernaak, University of Twente, Enschede, The Netherlands
A. Rantzer, Lund Institute of Technology, Sweden
J.N. Tsitsiklis, MIT, Cambridge, MA, USA

About this Series

This series aims to report new developments in the fields of control and information sciences—quickly, informally and at a high level. The type of material considered for publication includes:

1. Preliminary drafts of monographs and advanced textbooks
2. Lectures on a new field, or presenting a new angle on a classical field
3. Research reports
4. Reports of meetings, provided they are
   (a) of exceptional interest and
   (b) devoted to a specific topic. The timeliness of subject material is very important.

More information about this series at http://www.springer.com/series/642

Fabio Fagnani · Paolo Frasca

Introduction to Averaging Dynamics over Networks

Springer

Fabio Fagnani
Dipartimento di Scienze Matematiche
"Giuseppe Luigi Lagrange"
Politecnico di Torino
Turin
Italy

Paolo Frasca
GIPSA-lab
CNRS
Grenoble
France

ISSN 0170-8643 ISSN 1610-7411 (electronic)
Lecture Notes in Control and Information Sciences
ISBN 978-3-319-68021-7 ISBN 978-3-319-68022-4 (eBook)
https://doi.org/10.1007/978-3-319-68022-4

Library of Congress Control Number: 2017952924

Mathematics Subject Classification (2010): 15B51, 9302, 91D30, 15A18

© Springer International Publishing AG 2018
This work is subject to copyright. All rights are reserved by the Publisher, whether the whole or part of the material is concerned, specifically the rights of translation, reprinting, reuse of illustrations, recitation, broadcasting, reproduction on microfilms or in any other physical way, and transmission or information storage and retrieval, electronic adaptation, computer software, or by similar or dissimilar methodology now known or hereafter developed.
The use of general descriptive names, registered names, trademarks, service marks, etc. in this publication does not imply, even in the absence of a specific statement, that such names are exempt from the relevant protective laws and regulations and therefore free for general use.
The publisher, the authors and the editors are safe to assume that the advice and information in this book are believed to be true and accurate at the date of publication. Neither the publisher nor the authors or the editors give a warranty, express or implied, with respect to the material contained herein or for any errors or omissions that may have been made. The publisher remains neutral with regard to jurisdictional claims in published maps and institutional affiliations.

Printed on acid-free paper

This Springer imprint is published by Springer Nature
The registered company is Springer International Publishing AG
The registered company address is: Gewerbestrasse 11, 6330 Cham, Switzerland

Preface

This book has been written to be an agile but accurate introduction to the dynamics over network usually known as *averaging* or *consensus dynamics*. Mathematically, the simplest averaging dynamics is a linear dynamical system driven by a fixed stochastic matrix, whose zero pattern is determined by the network topology. The main goal of this book is proposing a unified and self-contained theoretical framework that is suitable to analyze not only this simple instance, but also several related dynamical models that feature time-variance, randomness, and heterogeneities. Even if most emphasis will be put on the methodological and general aspects of the subject, we will also treat applications to inferential problems in sensor networks, rendezvous of mobile robots, and opinion dynamics in social networks.

This book originated from our lectures for the graduate courses *Control of/over networks* and *Dynamics over networks*, taught at the Politecnico di Torino since 2011. The treatment is completely self-contained, with only standard linear algebra, calculus, and probability as prerequisites. For this reason and for the abundance of exercises, we hope that this text can be effectively used as a resource for teaching and self-study at graduate or advanced undergraduate level.

This preface has for us a threefold goal: concisely introducing the topic of this book, explaining our perspective in writing it, and outlining its contents.

Averaging Dynamics and Multi-agent Systems

Multi-agent systems constitute one of the fundamental paradigms of science and technology in the present century. Their key feature is that complex dynamical evolutions originate from the interactions of a large number of simple units. Not only such collective behaviors are evident in biological and social systems, but the digital revolution and the miniaturization in electronics have also made possible the creation of man-made complex architectures of interconnected devices, including computers, sensors, and cameras. Moreover, the creation of the Internet has enabled totally new forms of social and economic aggregation. These technological and social evolutions have strongly pushed researchers toward a deeper and more systematic study of multi-agent dynamical systems.

The mathematical structure of multi-agent systems is that of a graph (typically of large scale), where the nodes are agents or units endowed with some (typically simple) dynamical system. These dynamical systems are coupled through the edges of the graph. Complexity is thus the outcome of the topology and the nature of the interconnections, which may often be of stochastic nature. The typical mathematical issue is understanding how the topology of the graph affects the transient and asymptotic behavior of the intercoupled dynamics, relating graph-theoretical concepts (such as diameter, degrees, connectivity, presence of bottlenecks) to the dynamic behavior in a quantitative way.

In the applications, these dynamical systems can represent a multitude of different situations. For instance, the graph can be an infrastructure network (e.g., sensor or computer network) and the dynamics be an algorithm designed to fuse information and eventually reach a preassigned goal, such as estimation or synchronization, through cooperation. In other situations, the network may represent relationships between socioeconomic or financial units (people, companies) and the state variables represent opinions or other economic indicators. Finally, the units may be physically positioned in the space (such as animals, pedestrians, or vehicles) and have as state variables their positions and velocities: The dynamics then represent some collective motion, such as platooning for automated vehicles, formation flight for drones, or flocking for animals.

Averaging dynamics is one of the most popular and maybe the simplest multi-agent dynamics. It may be convenient to introduce it by the language of social sciences. Suppose that a number of individuals possess some information represented by a real number: For instance, such numbers can represent their opinions on a certain matter. The individuals interact and change their opinions by averaging them with the opinions of other individuals to which they are connected. Under certain assumptions, these updates will lead all the community to converge to a common opinion that depends on the initial opinions of all individuals. Because of its intrinsic push toward consensus, the averaging dynamics is also known as consensus dynamics.

Aims and Scope

While teaching this topic and preparing this manuscript, we have tried to follow four guidelines.

(i) We concentrate on *linear discrete-time averaging dynamics*, which we identify as the core theoretical issue. We present the fundamental results on averaging dynamics and a unified viewpoint of various models and results scattered in the literature. Starting from the classical evolution of the powers of a fixed stochastic matrix, we then consider more general products of a sequence of stochastic matrices.

(ii) We keep the discourse *simple and self-contained*. The necessary theory is constructed in this book without, in particular, assuming any knowledge of Markov chains or of Perron–Frobenius theory. All convergence results presented are proposed as derivations of two different principles. The first one is a "contraction" principle, prescribing that the convex hull of the states shrinks as time elapses. This principle applies to time-invariant dynamics and to time-varying dynamics where information is able to flow between any two units within a bounded time. The second principle, instead, postulates a "reciprocity" in the network dynamics, namely that if information can flow from a set of units to another, also the converse must be possible. This second principle also permits to prove convergence in settings where consensus is not necessarily reached.

(iii) We constantly aim to relate the properties of the information flow (essentially determined by a suitable graph) with the properties of dynamics. Dynamical properties of interest include not only mere convergence but also "performance," broadly intended. We indeed consider different notions of performance, including the rate of convergence, the accuracy in approximating the average of the initial states, and the robustness against noise and communication errors. In all these cases, the *relation between graph and performance* is made explicit by the spectral analysis of the update matrix.

(iv) We develop our approach in the perspective of *large-scale networks*: Even though our theory is valid for networks of any size, we pay special attention to how dynamical properties depend on the size of the network. Concretely, this leads us to specialize our results to specific families of graphs (for instance, grids) and to take limits where the number of nodes grows to infinity.

Contents

We now outline the contents of this book, even though the reader will be provided with more detailed summaries at the beginning of each of its five chapters.

The initial Chap. 1 presents the graph-theoretical background that is essential to our work, with particular emphasis on connectivity properties and on algebraic graph theory. We also define several graphs that are used as leading examples in the rest of this book.

Chapter 2 is the core of this book and is entirely devoted to time-invariant averaging dynamics. Actually, we jointly study the convergence of averaging dynamics and the properties of stochastic matrices, by means of a basic and fundamental contraction principle. We also pay attention to reversible matrices and present the classical Cheeger bound to estimate the second eigenvalue.

In Chap. 3, we study time-varying dynamics; that is, we allow the graph to change with time. We present an extension to the contraction principle already used in Chap. 2 and then propose a different convergence analysis based on a principle of reciprocity in the flow of information. A number of applications of these general results are presented including models where time-variance is due to a random mechanism, e.g., gossiping, and models exhibiting nonconsensus behaviors, e.g., bounded-confidence opinion dynamics in social networks.

Chapter 4 is devoted to a finer analysis of the time-invariant averaging dynamics. We define various performance metrics that quantify, for instance, convergence speed and robustness with respect to noise. Performance metrics are then evaluated as functions of the eigenvalues of the graph. We also present the application to distributed inferential estimation.

Finally, in Chap. 5, we develop the theory of electrical networks of resistors, which has important connections with reversible stochastic matrices. Using just linear algebra techniques and no probability, we present some basic concepts like that of voltage, Green matrix, harmonic extension, and effective resistance. This machinery is then used to address two different problems: (i) estimation from relative measurements and (ii) averaging dynamics in the presence of stubborn agents.

The main dependences between these chapters are straightforward to describe by a graph.

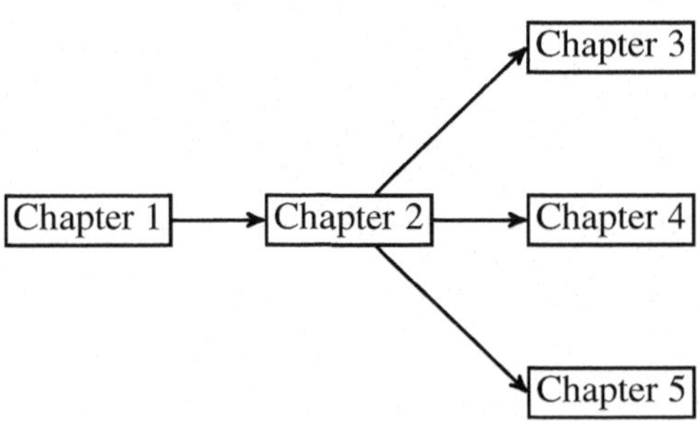

All chapters include extended examples and are concluded by a selection of Exercises (about 100 in total) and by some Bibliographical Notes, which have no ambition to be exhaustive but simply to provide some context and propose some additional readings.

Last but not least, we would like to acknowledge that our perspective on these topics has been shaped by fruitful collaborations with students and colleagues: A nonexhaustive list includes Ruggero Carli, Giacomo Como, Jean-Charles Delvenne, Federica Garin, Julien Hendrickx, Chiara Ravazzi, Wilbert Samuel Rossi, and Sandro Zampieri.

Turin, Italy
Grenoble, France
July 2017

Fabio Fagnani
Paolo Frasca

Contents

Chapter 1
Graph Theory

Abstract This chapter is a self-contained and concise introduction to graph theory, which is essential to study the averaging dynamics over networks. After some basic notions in Sect. 1.1, the emphasis is on connectivity and periodicity properties, which are presented in Sects. 1.2 and 1.3, respectively. Section 1.4 introduces the adjacency and Laplacian matrices associated to a given graph and studies their spectra. Finally, Sect. 1.5 introduces some notable examples of graphs, such as circulant, Cayley, and De Bruijn graphs.

1.1 Basic Definitions and Examples

We begin with the definition of graph, which is central in our studies. A graph G is a pair (V, E) where V is a finite set, whose elements are said to be the *vertices* (or *nodes*) of G, and $E \subset V \times V$ is the set of *edges* (or *arcs*). The cardinality of V is said to be the *order* or the *size* of the graph. An edge of the form (u, u) is said to be a *self-loop*, or simply a *loop*. In a graph, every arc represents a connection or link between two nodes. It is customary to draw graphs by representing nodes as dots and arcs as arrows connecting the nodes in such away that for an edge $(u, v) \in V \times V$, we understand that u is the tail and v is the head of the arrow; see Fig. 1.1. When drawing a graph, we are thus implicitly assigning a location in the plane to each node. The trivial graph $E_V = (V, \emptyset)$ is said to be an *empty* graph. On the opposite extreme, the graph $K_V = (V, \{(u, v) : u \neq v\})$ is said to be a *complete* graph (note that self-loops have been excluded, see Fig. 1.1). Two graphs $G = (V, E)$ and $G' = (V', E')$ are said to be *isomorphic* if there exists a bijection $\psi : V \to V'$ such that

$$(v, w) \in E \;\Leftrightarrow\; (\psi(v), \psi(w)) \in E'.$$

For instance, two complete graphs are isomorphic when they have the same order. For this reason, we may also denote a complete graph of order n simply as K_n. Essentially, two isomorphic graphs simply differ by a different labeling of the vertices. Since in all the applications we will consider such differences will not play any role, we will consider two isomorphic graphs as identical in what follows. This equivalence also

© Springer International Publishing AG 2018

F. Fagnani and P. Frasca, *Introduction to Averaging Dynamics over Networks*, Lecture Notes in Control and Information Sciences 472,
https://doi.org/10.1007/978-3-319-68022-4_1

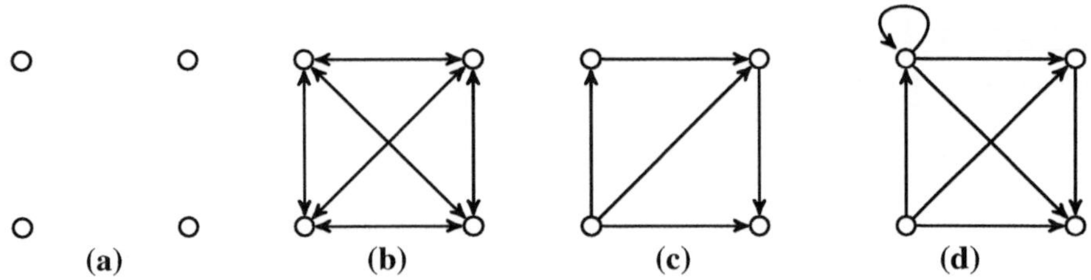

Fig. 1.1 Examples of graphs with four nodes: empty graph, complete graph, a graph without self-loops, and a graph with a self-loop

allows us to identify (when convenient) the vertex set with a set of numbers, writing for instance $V = \{1, \dots, |V|\}$.

Given a graph $G = (V, E)$, the *reverse* graph of G is the graph which is obtained by reversing all arcs. That is, $\mathrm{rev}(G) = (V, \{(u, v) \in V \times V : (v, u) \in E\})$. The special case in which $\mathrm{rev}(G) = G$ is very important, as it means that $(u, v) \in E$ if and only if $(v, u) \in E$. If a graph is such, it is said to be *symmetric*. When drawing a symmetric graph, there is no need to use pairs of arrows to connect nodes: In this case, we will rather use double-headed arrows or just segments; see Figs. 1.1 and 1.3.

If $(u, v) \in E$, then v is said to be a *out-neighbor* of u, and we write that $v \in \mathcal{N}_u^{\mathrm{out}}$. Conversely, u is said to be a *in-neighbor* of v in the graph, and we write $u \in \mathcal{N}_v^{\mathrm{in}}$. The number of out-neighbors of a node v is said to be its out-degree and is denoted by d_v^{out}. Correspondingly, the number of in-neighbors of a node v is said to be its in-degree and is denoted by d_v^{in}. Note that for every graph, the following identity holds true:

$$|E| = \sum_{v \in V} d_v^{\mathrm{out}} = \sum_{w \in V} d_w^{\mathrm{in}}. \tag{1.1}$$

A *source* is a node with no in-neighbors, and a *sink* is a node u with no out-neighbors. A graph is said to be d-(in/out-)regular if the (in/out-)degree of every node is d. A graph is *topologically balanced* if $d_v^{\mathrm{out}} = d_v^{\mathrm{in}}$ for all nodes v. Note that for a symmetric graph, there is no need to distinguish between in- and out-neighbors, so that any symmetric graph is topologically balanced and we will just talk about neighbors and degrees dropping the labels "in" and "out".

It is sometimes useful to identify certain relationships between graphs. The *intersection* and *union* of two graphs $G = (V, E)$ and $G' = (V', E')$ are denoted by, respectively, $G \cap G' = (V \cap V', E \cap E')$ and $G \cup G' = (V \cup V', E \cup E')$. On the other hand, we say that a graph $G' = (V', E')$ is a *subgraph* of $G = (V, E)$ if $V' \subset V$ and $E' \subset E$: this relation is denoted as $G' \subset G$. Furthermore, the subgraph G' is said to be *spanning* if $V' = V$ and is said to be the subgraph *induced* by V' if

$$E' = E \cap (V' \times V').$$

Fig. 1.2 Directed and undirected line graphs on four nodes

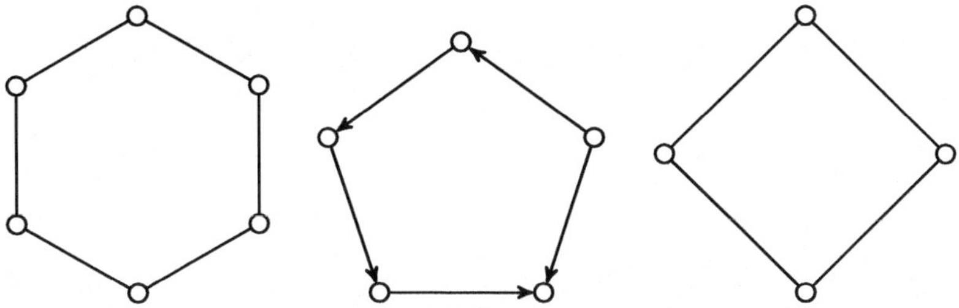

(a) C_6 (b) $\vec{C}_5$ (c) C_4

Fig. 1.3 Three examples of directed and undirected cycle graphs of small size

In this last case, we use the notation $G' = G_{|V'}$. Clearly, any graph on the vertex set V, without self-loops, is a subgraph of the complete graph K_V.

Next, we provide some more examples of graphs. Let $V = \{0, \ldots, n-1\}$.

(i) If $E = \{(u, v) \in V \times V : |v - u| = 1\}$, then (V, E) is said to be a *line* graph and is denoted as L_n.
(ii) If $E = \{(u, v) \in V \times V : v - u = 1\}$, then (V, E) is said to be a *directed line* graph and is denoted as $\vec{L}_n$.
(iii) If $E = \{(u, v) \in V \times V : v - u = 1 \mod n\}$, then (V, E) is said to be a *directed cycle* graph and is denoted as $\vec{C}_n$.
(iv) If $E = \{(u, v) \in V \times V : (v - u) \mod n \in \{-1, +1\}\}$, then (V, E) is said to be a *cycle* graph and is denoted as C_n.

Some properties of line and cycle graphs can be immediately observed: For example, $\vec{L}_n \subset L_n$, and more precisely, $L_n = \vec{L}_n \cup \text{rev}(\vec{L}_n)$. Correspondingly, $\vec{C}_n \subset C_n$ and $C_n = \vec{C}_n \cup \text{rev}(\vec{C}_n)$. Moreover, the cycle graph C_n is 2-regular and the directed cycle graph $\vec{C}_n$ is topologically balanced. Examples of cycle and line graphs are drawn in Figs. 1.2 and 1.3.

1.2 Paths and Connectivity

In this section, we turn our attention to investigate the connectivity properties of graphs. The pictorial representation of graphs, which we have introduced above, makes the following definitions very natural.

Given a graph $G = (V, E)$ and a pair of nodes u, v, a *path* (of *length* l) from u to v on G is an ordered list of nodes $(w_0, \ldots, w_l)$ such that

(i) $w_0 = u$ and $w_l = v$;
(ii) $(w_i, w_{i+1}) \in E$ for every $i \in \{0, \ldots, l-1\}$.

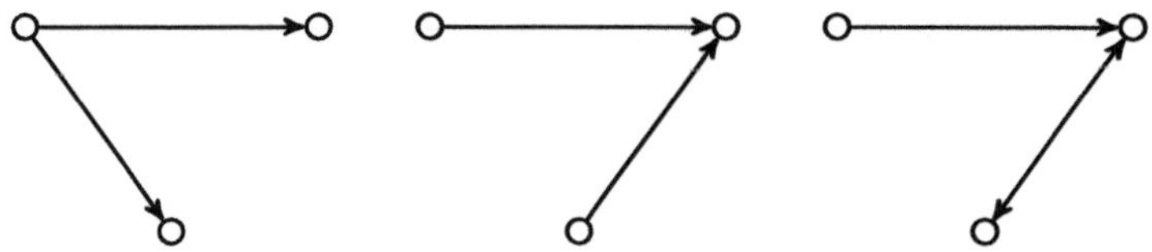

Fig. 1.4 Connectivity examples using graphs with three nodes: (weakly) connected without a globally reachable node, weakly connected with a globally reachable node, and strongly connected

The edges occurring in the definition of a path are said to *insist* on the path at hand. The path is said to be *simple* if the edges (w_i, w_{i+1}) are all distinct. If a path from u to v exists, we say that v is *reachable* from u. Given two nodes u and v, we say that they *communicate* if either $u = v$ or $u \neq v$ and there are both a path from u to v and one from v to u. It is easy to check—Exercise 1.2—that communication is an equivalence relation between nodes. These notions are instrumental to the following important definitions.

A graph $G = (V, E)$ is said to be

- *strongly connected* if every two nodes communicate;
- *connected* if for any pair of nodes (u, v), either u is reachable from v or v is reachable from u;
- *weakly connected* if $G \cup \text{rev}(G)$ is strongly connected.

Note that these three definitions are equivalent for symmetric graphs. We also note that every graph can be seen as the disjoint union of weakly connected subgraphs, which we call *weakly connected components* or simply *connected components*.

A node v is said to be *globally reachable* if for every other node w there exists a path from w to v. Clearly, in a strongly connected graph, all nodes are globally reachable. A partial converse is given by the following result:

Proposition 1.1 (Connectivity and balance) *If G is topologically balanced and contains a globally reachable node, then G is strongly connected.*

Proof By contradiction, the graph $G = (V, E)$ is not strongly connected. Let R be the set of globally reachable nodes: By the assumptions, $\emptyset \subsetneq R \subsetneq V$. Consider the partition of nodes into R and $V \setminus R$, and note that there is no edge from R to $V \setminus R$ but there is at least one edge from $V \setminus R$ to R. Let v be the tail of such edge. By the balance property, there must be an edge (u, v) with $u \notin R$. In turns, the same remark implies that there exists an edge (t, u) with $t \notin R$. As the set $V \setminus R$ is finite, this iterative procedure must end after a finite number of steps, showing that there is at least one node in $V \setminus R$ that has different in-degree and out-degree, contradiction.□

Some examples of graph connectivity are given in Fig. 1.4.

The notion of path is also the ground to endow graphs with a natural distance between nodes. As we have defined above, the *length* of a path is the number of edges insisting on the path. Then, given two nodes u and v of a graph $G = (V, E)$, we can define the *distance from u to v* as the length of the shortest path which connects them. Precisely, we let[1]

$$\mathrm{dst}_G(u, v) = \min\{\ell : \text{ there exists in } G \text{ a path of length } \ell \text{ from } u \text{ to } v\}, \qquad (1.2)$$

provided $u \neq v$, and $\mathrm{dst}_G(u, u) = 0$. Note that the function $\mathrm{dst}_G(\cdot, \cdot)$ is symmetric in its arguments if G is a symmetric graph (cf. also Exercise 1.11). Furthermore, the *diameter* of the graph $G = (V, E)$ is defined as

$$\mathrm{diam}(G) = \max\{\mathrm{dst}_G(u, v) \ : \ u, v \in V\}.$$

Clearly, G is strongly connected if and only if $\mathrm{diam}(G)$ is finite. Moreover, for any strongly connected graph G of order n, it holds that $\mathrm{diam}(G) \leq n - 1$. It is easy to compute the diameter for the graph examples introduced above: For instance, for every $n \in \mathbb{N}$ we have $\mathrm{diam}(K_n) = 1$ and $\mathrm{diam}(C_n) = \lfloor n/2 \rfloor$.

A very important class of paths are "closed" paths: A path from a node to itself is said to be a *circuit*. For instance, loops are circuits of length one. A graph is said to be *circuit-free* if it contains no circuit. The following is a simple property of circuit-free graphs.

Proposition 1.2 (Source and sink) *Every circuit-free graph has at least one source and at least one sink.*

Proof By contradiction, we take a graph $G = (V, E)$ with no sink, that is such that $d_u^{\mathrm{out}} \geq 1$ for every $u \in V$. We pick any vertex and denote it as v_0. Then, we take one out-neighbor of v_0 and denote it by v_1. Then, recursively for $k \geq 1$, we take v_{k+1} among the out-neighbors of v_k. As the cardinality of V is finite, it must happen for a certain $\ell \in \mathbb{N}$ that $v_{\ell+1}$ belongs to $\{v_0, \dots, v_\ell\}$, thus forming a circuit and providing the required contradiction. The existence of a source is proven similarly. □

When a graph $G = (V, E)$ is not strongly connected, we can consider its *strongly connected components*, which we define as follows. First, we have observed—see Exercise 1.2—that the relation of communication between nodes is an *equivalence relation*. Then, we can consider the partition of V into the corresponding equivalence classes $V = V_1 \cup V_2 \cup \cdots \cup V_s$ and the induced subgraphs $G_i = \big(V_i, E \cap (V_i \times V_i)\big)$, which are called the strongly connected components of G. If the graph G is symmetric, actually G is simply the union of these s subgraphs, in the sense that there is no further edge in the graph, the connected components being completely isolated from each other. For general graphs, the situation is more complicated: A useful way to describe what is left beyond the strongly connected components is the following concept of *condensation graph*, whose nodes represent the strongly connected components of G.

Definition 1.1 (*Condensation graph*) Given any graph $G = (V, E)$, consider its strongly connected components $G_k = (V_k, E_k)$, $k \in \{1, \dots, s\}$. The condensation

[1] We understand that the minimum of an empty set is $+\infty$.

Fig. 1.5 An example of graph G (*left*) with its condensation graph $\mathscr{T}(G)$ (*right*)

graph of G is a graph $\mathscr{T}(G)$ with set of vertices $\{1, \ldots, s\}$ such that there is an arc in $\mathscr{T}(G)$ from h to k if $k \neq h$ and there is an arc in G from a vertex in V_k to a vertex in V_h.

The construction is illustrated in Fig. 1.5. We leave to the reader the task of proving the following properties of condensation graphs.

Proposition 1.3 (Condensation graphs) *Let G be any graph and $\mathscr{T}(G)$ its condensation graph. Then,*

(i) *$\mathscr{T}(G)$ is circuit-free;*
(ii) *$\mathscr{T}(G)$ is (weakly) connected if and only if G is (weakly) connected;*
(iii) *G contains a globally reachable node if and only if $\mathscr{T}(G)$ has only one sink.*

A *cycle* is a circuit of length at least 3, with no vertex repeated except the first and last one. A graph is said to be *cycle-free* if it contains no cycles, and *unicyclic* if it contains exactly one cycle. A *tree* is a symmetric cycle-free connected graph, and an cycle-free symmetric graph is also called a *forest*. The next result states some relevant properties of trees: other properties are presented in Exercise 1.5.

Proposition 1.4 (Trees) *Let $G = (V, E)$ be a symmetric graph. Then, the following four statements are equivalent.*

(i) *G is a tree;*
(ii) *for any pair of distinct nodes u and v in V, there is exactly one path from u to v in G;*
(iii) *G is* minimal connected, *that is, G is strongly connected and removing any edge makes the resulting graph not strongly connected;*
(iv) *G is* maximal cycle-free, *that is, G is cycle-free and adding one edge creates a cycle in G.*

Proof The key point of this proof is the equivalence between (i) and (ii). Indeed, assume G is a tree, that is, G is connected symmetric and cycle-free. Then, G is strongly connected, and thus, there is a path connecting u to v. Furthermore, if there was another path, the graph being symmetric would imply the existence of a cycle. Conversely, the existence of exactly one path implies connectedness and absence of cycles. Next, we can observe that removing any edge necessarily breaks at least one path, thus causing a graph satisfying (ii) to become not strongly connected. Conversely, property (iii) ensures that there are no multiple paths connecting the nodes, for otherwise strong connectivity would be robust to edge deletions. As we have noted that in symmetric graphs the absence of multiple paths is equivalent to the absence of cycle, we conclude that (iii) implies (ii). Proving the equivalence between (iii) and (iv) is left to the reader. □

1.3 Periodicity

Given a graph $G = (V, E)$ and $v \in V$, denote by L_v the set of lengths of the circuits in G to which v belongs. The period of v is the greatest common divisor (GCD) of the integers in L_v (if $L_v = \emptyset$, the period is undefined). The node is said to be *aperiodic* if its period is one. Notice that if a self-loop $(v, v) \in E$ is present, then $1 \in L_v$ and v is thus certainly aperiodic. The graph itself is said to be *aperiodic* if every node is aperiodic.

Example 1.1 In the directed cycle graph $\vec{C}_n$, each node has period equal to n. In the symmetric cycle graph C_n, instead, the period of each node is equal to GCD$(2, n)$. In particular, symmetric cycle graphs C_n with n odd are all aperiodic.

Notice that, since circuits can be concatenated freely to obtain new circuits, it follows that the length sets L_v are closed under addition[2] ($\ell_1, \ell_2 \in L_v$ yield $\ell_1 + \ell_2 \in L_v$). For aperiodic nodes, something very strong can be stated about L_v. We start recalling the following well-known fact from algebra.

Lemma 1.1 (Bézout's identity) *Let $a_1, \ldots, a_s \in \mathbb{N}$ and let $d \in \mathbb{N}$ be their GCD. Then, there exist s coefficients $\alpha_i \in \mathbb{Z}$ such that $\sum_i \alpha_i a_i = d$.*

By this lemma, we can prove the following key result.

Proposition 1.5 (Aperiodicity) *Let $G = (V, E)$ be a graph and let $v \in V$. The following conditions are equivalent.*

(i) v is aperiodic;
(ii) there exists $m \in \mathbb{N}$ such that $m, m + 1 \in L_v$;
(iii) there exists $\ell \in \mathbb{N}$ such that for every $n \geq \ell$ it holds that $n \in L_v$.

Proof Clearly, (iii) $\Rightarrow$ (ii) $\Rightarrow$ (i).

(i) $\Rightarrow$ (ii): Since v is aperiodic, we can find lengths $\ell_1, \ell_2, \ldots, \ell_s \in L_v$ such that $1 = \text{GCD}(\ell_1, \ldots, \ell_s)$. Hence, by Lemma 1.1, we can find numbers $\alpha_i \in \mathbb{Z}$ such that $1 = \sum_{i=1}^{s} \alpha_i \ell_i$. Let $m = \sum_{i=1}^{s} |\alpha_i| \ell_i$ and notice that $m + 1 = \sum_{i=1}^{s} (|\alpha_i| + \alpha_i) \ell_i$. This shows that both $m, m + 1 \in L_v$, yielding (ii).

(ii) $\Rightarrow$ (iii): Notice first that if $m = 1$ in (ii), then (iii) is immediate. Suppose now that $m > 1$ and put $\ell = (m - 1)m$. Let $n \geq \ell$. Dividing n by m, we obtain

[2] Note that this also implies closure under integer multiplication as $\ell \in L_v$ and $m \in \mathbb{N}$ yield $m\ell \in L_v$.

$n = mh + r = m(h - r) + (m + 1)r$. Since, by definition of rest, $r \leq m - 1$ and by our choice of ℓ the quotient satisfies $h \geq m - 1$, we have that $h - r \geq 0$. Since both m and $m + 1$ belong to L_v, the last inequality implies that $n \in L_v$. □

The result above shows that for every aperiodic node, there exists $\ell \in \mathbb{N}$ such that for every $n \geq \ell$ there exists a path of length n from the node to itself. In other words, for every aperiodic node, there exist paths of *any* length from the node to itself—possibly excluding lengths below a certain threshold. This fact clearly means a great "freedom of movement" in the graph. Furthermore, aperiodicity of a single node is easily inherited by the rest of the graph, as a consequence of the following result.

Proposition 1.6 (Aperiodic vertices) *Let $G = (V, E)$ be a graph and let $u, v \in V$ be two communicating nodes. Then, u is aperiodic if and only if v is aperiodic.*

Proof As the node u is aperiodic, there exists $\ell \in \mathbb{N}$ and two circuits from u to itself which have lengths ℓ and $\ell + 1$. By the communication assumption, there exist a path from u to v (of length m) and a path from v to u (of length n). Hence, there exist two circuits (possibly repeating vertices) from v to itself, which have lengths $m + \ell + n$ and $m + \ell + n + 1$, proving the thesis. □

As a corollary, if a graph is strongly connected and has an aperiodic vertex, the graph is aperiodic. The above discussion also allows us to conclude the following remarkable result.

Corollary 1.1 (Paths on strongly connected and aperiodic graphs) *If a graph $G = (V, E)$ is strongly connected and aperiodic, then there exists ℓ such that for any pair of nodes u, v and any length $m \geq \ell$ there is a path from u to v of length m.*

Next, we present an additional notion which relates to paths and connectivity. A graph (V, E) is said to be *bipartite* if the set V can be apportioned into two subsets V_1 and V_2 such that for all $(u, v) \in E$ either $u \in V_1$ and $v \in V_2$ or $u \in V_2$ and $v \in V_1$. We already know some examples of bipartite graphs. For instance, trees are bipartite and C_n is bipartite if and only if n is even. The following is another natural example.

Example 1.2 (*Complete bipartite*) Let A, B be two nonempty sets of cardinalities m and n, respectively. The *complete bipartite* graph $K_{m,n}$ is the graph with node set $A \cup B$ and an edge (u, v) if and only if $u \in A$ and $v \in B$ or $u \in B$ and $v \in A$.

An important characterization of bipartite graphs is given by the following result.

Proposition 1.7 (Bipartition condition) *A graph is bipartite if and only if every circuit has even length.*

Proof If the graph $G = (V, E)$ is bipartite with $V = V_1 \cup V_2$, every path $u_0, u_1, \ldots, u_p$ having $u_0 \in V_1$ is such that $u_i \in V_1$ if and only if i is even. Therefore, if $u_p = u_0$, then p must be even. In order to prove the converse statement, we construct the partition $\{V_1, V_2\}$. We take any circuit $u_0, u_1, \ldots, u_{p-1}, u_0$ and let u_i belong to

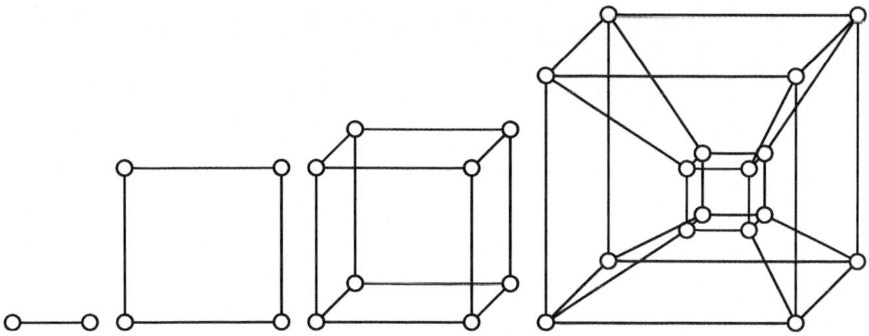

(a) H_1 **(b)** H_2 **(c)** H_3 **(d)** H_4

Fig. 1.6 Hypercubes of dimensions until four

V_1 if i is even, and to V_2 otherwise. Note that there is no edge connecting u_i and u_j when i and j are both even (or both odd), for otherwise the "shortcut" would create a circuit with odd length. Next, we move to another circuit $w_0, \ldots, w_q$. If the circuit has no intersection with that examined before, we can just repeat the same reasoning. If instead, say, the new circuit contains a node s which has already been attributed to V_1, we denote $s = w_0$ and proceed as above. Note that the procedure can be performed without introducing contradictions, because of the absence of circuits of odd length. Iterating the procedure constructs the required partition. □

As a corollary, we note that any bipartite graph is not aperiodic.

Next, we introduce a remarkable family of graphs, which the reader may easily verify to be bipartite.

Example 1.3 (*Hypercube*) Let V be the set of the binary words of length n, that is, $V = \{0, 1\}^n$. Then, the hypercube H_n is the graph on V with an edge between two words whenever they differ in exactly one component, i.e., $E = \{(u, v) : \|u - v\|_1 = 1\}$. It is immediate to observe that $|V| = 2^n$, and that H_n is symmetric, n-regular, and bipartite. Hypercube graphs are so-called because they draw the vertices and edges of n-dimensional cubes: This can be observed from the examples in Fig. 1.6.

1.4 Matrices and Eigenvalues

This section introduces (i) relevant matrices which are used in the study of graphs, namely the adjacency and Laplacian matrices; (ii) the notion of graph associated with a matrix; and (iii) the definition of weighted graph that inherently involves a matrix. Relating graphs to matrices permits to take advantage of algebraic tools for the study of graphs, and conversely to express matrix properties in terms of graphs.

Furthermore, the spectrum of adjacency and Laplacian matrices conveys important information about the graphs: This study is the topic of spectral graph theory.

First, we provide the fundamental definition of *adjacency matrix*. Given a graph $G = (V, E)$, the adjacency matrix A (sometimes denoted as A_G) is a matrix in $\{0, 1\}^{V \times V}$ such that

$$\begin{cases} A_{uv} = 1 & \text{if } (u, v) \in E \\ A_{uv} = 0 & \text{if } (u, v) \notin E. \end{cases}$$

As an example, observe that the adjacency matrix of the third graph in Fig. 1.1 is

$$A = \begin{bmatrix} 0 & 1 & 0 & 0 \\ 0 & 0 & 1 & 0 \\ 0 & 0 & 0 & 0 \\ 1 & 1 & 1 & 0 \end{bmatrix}.$$

The adjacency matrix encodes all the information about the structure of the graph and is thus a very important notion. Furthermore, it permits to answer questions about paths and connectivity of the graph by purely algebraic computations. The next result is a chief example.

Proposition 1.8 (Adjacency matrix and paths) *Let $G = (V, E)$ be a graph with adjacency matrix A. Then, for all $u, v \in V$ and $k \in \mathbb{N}$, the u, v-entry of A^k equals the number of paths of length k (including paths with self-loops) from node u to node v.*

Proof The statement is proved by induction on the length k. By definition of adjacency matrix, the statement is true for $k = 1$. Next, we assume that the statement is true for k and we prove it for $k + 1$. Note that each path from u to v of length $k + 1$ consists of an edge (u, w) and a path from w to v of length k. Since we can write

$$(A^{k+1})_{uv} = \sum_{w \in V} A_{uw} (A^k)_{wv},$$

the statement follows by the inductive hypothesis. □

Proposition 1.8 and Corollary 1.1 imply the following fact.

Corollary 1.2 (Adjacency matrix of aperiodic graphs) *If a graph $G = (V, E)$ is strongly connected and aperiodic, then there exists $\ell \in \mathbb{N}$ such that, for every $m \geq \ell$, every entry of A^m is strictly positive.*

By writing the adjacency matrix, we associate a matrix to any graph: Conversely, we may associate a graph to any matrix. Let M be a square matrix with nonnegative entries, whose rows and columns are indexed in a set V. Then, the graph *associated* to M, denoted by $\mathcal{G}_M$, is the graph (V, E) such that

$$E = \{(u, v) \in V \times V \; : \; M_{uv} > 0\}.$$

$$M = \begin{bmatrix} 0.3 & 2 & 1 \\ 0 & 0 & 1 \\ 0 & 0 & 0 \end{bmatrix}$$

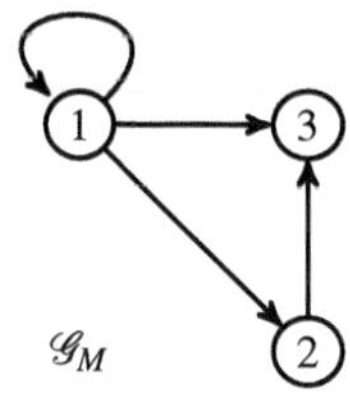

Fig. 1.7 A nonnegative matrix and its associated graph

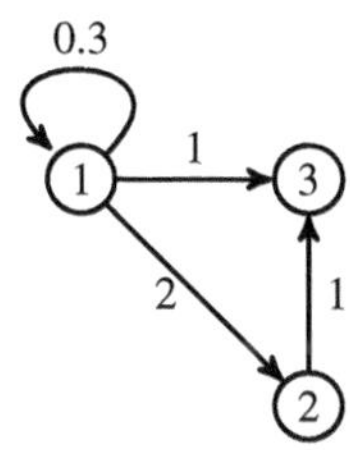

Fig. 1.8 The matrix M and the graph $\mathcal{G}_M = (V, E)$ in Fig. 1.7 drawn as a weighted graph (V, E, M)

An example is shown in Fig. 1.7. Based on this definition, we say that given a graph $G = (V, E)$, a matrix $M \in \mathbb{R}_{\geq 0}^{V \times V}$ is said to be *adapted to* G when $M_{uv} > 0$ and $u \neq v$ imply $(u, v) \in E$. Equivalently, we may say that $\mathcal{G}_M \subset G$ modulo self-loops. Of course, the adjacency matrix of a graph is an example of a matrix adapted to it.

Example 1.4 Consider the graph $\mathcal{G}_M$ in Fig. 1.7 and the matrices

$$M_1 \begin{bmatrix} 1 & 2.3 & 1 \\ 0 & 0 & 20 \\ 0 & 0 & 0 \end{bmatrix} \qquad M_2 = \begin{bmatrix} 1 & 0 & 1 \\ 0 & 0.1 & 2 \\ 0 & 0 & 0 \end{bmatrix} \qquad M_3 = \begin{bmatrix} 1 & 3.2 & 0 \\ 0.5 & 0 & 20 \\ 0 & 3 & 0 \end{bmatrix}.$$

Then, M_1 and M_2 are adapted to the graph, whereas M_3 is not.

Sometimes in the literature a graph $G = (V, E)$ together with a matrix $M \in \mathbb{R}_{\geq 0}^{V \times V}$, adapted to G, is called a *weighted graph* and is also denoted as (V, E, M). Weighted graphs are often depicted as graphs with each edge (v, w) labeled with M_{vw}, as in Fig. 1.8. Depending on the applications, M_{vw} may have a variety of different meanings: It can measure the capacity of a certain connection, a flow, a resistance, a distance, and so on.

We now propose the following fundamental construction. The Laplacian matrix associated with a matrix $M \in \mathbb{R}_{\geq 0}^{V \times V}$ is the matrix $L(M) \in \mathbb{R}^{V \times V}$ such that $L(M)_{uv} = -M_{uv}$ if $u \neq v$ and $L(M)_{uu} = \sum_{v:v \neq u} M_{uv}$. In matrix form, we may write that

$$L(M) = \text{diag}(M\mathbf{1}) - M,$$

where the notation $\text{diag}(x)$ denotes the square matrix whose main diagonal is the vector x. In the special case when M is the adjacency matrix of the graph G, the resulting Laplacian matrix $L(A_G)$ is simply called the *Laplacian* of G and denoted with the symbol L_G. Notice that in this case, L_G has the form

$$L_G = D_G - A_G$$

where D_G is a diagonal matrix such that $(D_G)_{uu} = d_u^{\text{out}}$ for every $u \in V$. As an immediate consequence, we observe that $L(M)$ does not depend on the diagonal values of M. In particular L_G is independent of the presence of self-loops in the graph. As well, it is immediate that the graph G is symmetric if and only if L_G is symmetric.

The spectrum of $L(M)$ plays an important role in graph theory and in many of the arguments which will be discussed later on. From the definition, it is immediate that for any matrix M, it holds that $L(M)\mathbf{1} = 0$, that is, 0 is an eigenvalue of $L(M)$ with eigenvector $\mathbf{1}$. We now propose a number of results in the case when M is symmetric. Possible extensions and refinements are outlined in Exercises 1.21 and 2.25. We start with the following basic fact.

Proposition 1.9 (Dirichlet form) *Let $M \in \mathbb{R}^{V\times V}$ be a symmetric matrix. For every $x \in \mathbb{R}^V$, it holds*

$$x^* L(M) x = \frac{1}{2}\sum_{u,v} M_{uv}(x_u - x_v)^2. \tag{1.3}$$

Proof By computing the quadratic form, we obtain

$$\begin{aligned}
x^* L(M) x &= \sum_{u}\sum_{v:v\neq u} M_{uv} x_u^2 - \sum_{u,v:u\neq v} M_{uv} x_u x_v \\
&= \sum_{u,v:u\neq v} M_{uv}(x_u^2 - x_u x_v) \\
&= \frac{1}{2}\left[\sum_{u,v:u\neq v} M_{uv}(x_u^2 - x_u x_v) + \sum_{u,v:u\neq v} M_{vu}(x_v^2 - x_u x_v)\right] \\
&= \frac{1}{2}\sum_{u,v:u\neq v} M_{uv}(x_u^2 - 2x_u x_v + x_v^2).
\end{aligned}$$

Notice the so-called "symmetrization" trick used in the third equality and the crucial role played by symmetry in the fourth equality. □

The previous result has a number of straightforward consequences.

Proposition 1.10 (Laplacian and connectivity) *Suppose M is symmetric. Then,*

(i) the Laplacian $L(M)$ is positive semidefinite;
(ii) the multiplicity of the eigenvalue 0 equals the number of connected components of $\mathcal{G}_M$.

Proof Exercise. Hint: to prove (ii), find suitable independent eigenvectors, each of them corresponding to a connected component. □

When M is symmetric of order n, eigenvalues of $L(M)$ are real and nonnegative: $0 = \lambda_1 \leq \lambda_2 \cdots \leq \lambda_n$. Particularly relevant is the "second eigenvalue" λ_2, which admits the following variational characterization.

Proposition 1.11 (Variational characterization) *If M is symmetric, then it holds*

$$\lambda_2 = \min_{x \neq 0, x^* \mathbf{1} = 0} \frac{x^* L(M) x}{x^* x}. \tag{1.4}$$

Proof Since $L(M)$ is symmetric we can find an orthonormal basis of eigenvectors: $x_{(i)} \in \mathbb{R}^V$ for $i \in \{1, \ldots, n\}$ with $L(M)x_{(i)} = \lambda_i x_{(i)}$. We can assume that $x_{(1)} = n^{-1/2}\mathbf{1}$. $L(M)$ can be expressed as a combination of orthogonal projectors:

$$L(M) = \sum_{i \geq 2} \lambda_i x_{(i)} x_{(i)}^*$$

(this corresponds to diagonalizing $L(M)$ with respect to the basis of eigenvectors). Hence, if $y \in \mathbb{R}^V$ is such that $\mathbf{1}^* y = 0$

$$y^* L(M) y = \sum_{i \geq 2} \lambda_i (x_{(i)}^* y)^2 \geq \lambda_2 \sum_{i \geq 2} (x_{(i)}^* y)^2 = \lambda_2 ||y||_2^2$$

This yields $\leq$ in (1.4). On the other hand, it holds

$$\frac{x_{(2)}^* L(M) x_{(2)}}{x_{(2)}^* x_{(2)}} = \lambda_2$$

and thus also $\geq$ is proven. □

This result has an immediate consequence.

Proposition 1.12 (Monotonicity) *Let M and Q be two symmetric matrices such that $M_{uv} \geq Q_{uv}$ for every $u \neq v$. If we denote by $\lambda_2(M)$ and $\lambda_2(Q)$ the second eigenvalues of, respectively, $L(M)$ and $L(Q)$, then $\lambda_2(M) \geq \lambda_2(Q)$.*

Proof It easily follows combining (1.4) with the quadratic form given by (1.3). □

1.5 Examples of Graphs

This section regards notable families of graphs and their properties. We also introduce a notion of product between graphs, which is useful to construct further examples.

1.5.1 Circulant Graphs

An $n \times n$ matrix A is said to be *circulant* if it exists a vector $c \in \mathbb{R}^n$ such that

$$A = \begin{bmatrix} c_0 & c_1 & c_2 & \dots & c_{n-1} \\ c_{n-1} & c_0 & c_1 & c_2 & \dots \\ & & \vdots & & \\ c_1 & \cdots & & c_{n-1} & c_0 \end{bmatrix}.$$

This matrix can be denoted as $A = \mathrm{circ}(c)$, where $c = [c_0, c_1, \dots, c_{n-1}]$. A graph is said to be *circulant* if its adjacency matrix is circulant. Examples include the directed cycle and cycle graphs introduced earlier.

Circulant matrices enjoy many interesting properties: Here we are interested in simple properties of their spectra, summarized in the following result.

Proposition 1.13 (Spectra of circulant matrices) *Let $c \in \mathbb{R}^n$ and consider the $n \times n$ circulant matrix $A = \mathrm{circ}(c)$. Then, A has eigenvectors*

$$x^{(k)} = [1, \omega_k, \dots, \omega_k^{n-1}]^\top \qquad k \in \{0, \dots, n-1\},$$

and corresponding eigenvalues $\lambda_k = \sum_{\ell=0}^{n-1} c_\ell \omega_k^\ell$, where we have denoted $\omega_k = e^{i \frac{2\pi}{n} k}$ with i the imaginary unit. Furthermore, the eigenvectors $x^{(k)}$ form an orthonormal basis of $\mathbb{C}^n$.

Proof Note that ω_k is such that $\omega_k^n = 1$: Indeed, ω_k is said to be a *n-th root of unity*. We leave to the reader to verify that the set of vectors $\{x^{(k)}\}_{k=0}^{n-1}$ is orthonormal, and we instead verify that the pair $(x^{(k)}, \lambda_k)$ satisfies the eigenvalue definition. To this goal, we compute

$$\begin{aligned} Ax^{(k)} &= c_0 \begin{bmatrix} 1 \\ \omega_k \\ \omega_k^2 \\ \vdots \\ \omega_k^{n-1} \end{bmatrix} + c_1 \begin{bmatrix} \omega_k \\ \omega_k^2 \\ \vdots \\ \omega_k^{n-1} \\ 1 \end{bmatrix} + \dots + c_{n-1} \begin{bmatrix} \omega_k^{n-1} \\ 1 \\ \omega_k \\ \omega_k^2 \\ \vdots \end{bmatrix} \\ &= \sum_{\ell=0}^{n-1} c_\ell \omega_k^\ell \begin{bmatrix} 1 \\ \omega_k \\ \omega_k^2 \\ \vdots \\ \omega_k^{n-1} \end{bmatrix} \\ &= \sum_{\ell=0}^{n-1} c_\ell \omega_k^\ell x^{(k)} \end{aligned}$$

thus proving the thesis. □

As special cases, we can compute the spectra of cycle graphs.

Example 1.5 (*Spectra of cycles*) The graph $\vec{C}_n$ has Laplacian eigenvalues

$$\lambda_k = 1 - \exp\left(i\frac{2\pi}{n}k\right) \qquad k \in \{0, \ldots, n-1\}.$$

The graph C_n has Laplacian eigenvalues

$$\lambda_k = 2 - 2\cos\left(\frac{2\pi}{n}k\right) \qquad k \in \{0, \ldots, n-1\}.$$

Note that these eigenvalues are real, $\lambda_k = \lambda_{n-k}$, and indeed a basis of real eigenvectors can be found (exercise).

1.5.2 Product Graphs

In this paragraph, we introduce the binary operations of *Cartesian product* between matrices and between graphs. Consider two matrices $A \in \mathbb{R}^{V\times V}$ and $B \in \mathbb{R}^{H\times H}$. We start recalling the familiar Kronecker product of matrices $A \otimes B \in \mathbb{R}^{(V\times H)\times(V\times H)}$ defined by

$$(A \otimes B)_{uh,vk} = A_{uv}B_{hk}. \tag{1.5}$$

The Cartesian product is instead defined as the matrix $A \times B \in \mathbb{R}^{(V\times H)\times(V\times H)}$ such that

$$(A \times B)_{uh,vk} = A_{uv}\delta_{hk} + B_{hk}\delta_{uv}, \tag{1.6}$$

where δ_{hk} is the standard Kronecker delta ($\delta_{hh} = 1$ and $\delta_{hk} = 0$ if $h \neq k$). This definition can be conveniently rewritten, by using the Kronecker product, as

$$A \times B = A \otimes I_H + I_V \otimes B\,, \tag{1.7}$$

where I_V and I_H are the identity matrices. Interpreting $\mathbb{R}^{V\times H}$ as the space of real matrices with rows and columns labeled by, respectively, elements of V and H, the Kronecker and Cartesian products can be thought as linear applications acting on matrices. Given a matrix $M \in \mathbb{R}^{V\times H}$, they can be equivalently expressed as

$$(A \otimes B)M = AMB^\top \quad \text{and} \quad (A \times B)M = AM + MB^\top. \tag{1.8}$$

Both the Kronecker and the Cartesian product are associative and commutative up to relabeling vertices (see also Exercise 1.26).

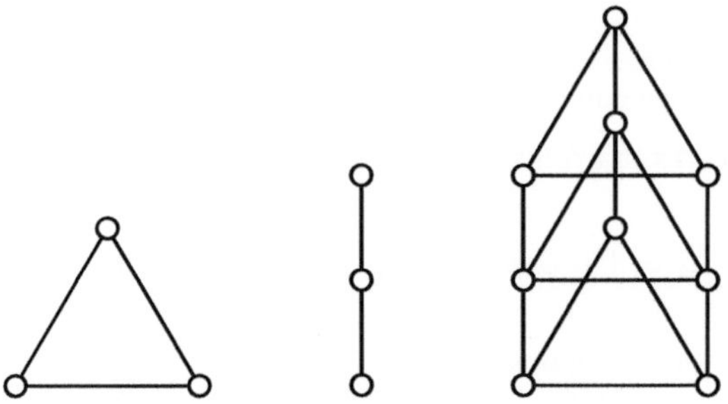

Fig. 1.9 Computing the Cartesian product between K_3 and L_3

The Cartesian product of matrices has a natural counterpart at the level of graphs. The (Cartesian) product of two graphs G and G' is the graph $G \times G'$, such that the vertex set is the Cartesian product of the vertex sets of G and G', and two vertices are adjacent when they agree in one coordinate and are adjacent in the other. A pictorial example is given in Fig. 1.9. We easily see that, as long as G and G' have no self-loops, the adjacency matrix of $G \times G'$ is $A_G \times A_{G'}$.

Furthermore, the degree of a node (u, h) in a product graph $G \times G'$ is the sum of the degrees of u in G and h in G'.

Proposition 1.14 (Product and Laplacian) *Let $A \in \mathbb{R}^{V\times V}$ and $B \in \mathbb{R}^{H\times H}$ be square matrices. Then,*

$$L(A \times B) = L(A) \times L(B).$$

Proof We start with a remark on products of diagonal operators. If $M \in \mathbb{R}^{V\times H}$, $\mathrm{diag}(M) \in \mathbb{R}^{(V\times H)\times(V\times H)}$ denotes the diagonal operator such that $\mathrm{diag}(M)_{uh,uh} = M_{uh}$. Given two vectors $x \in \mathbb{R}^V$ and $y \in \mathbb{R}^H$, it follows from definition (1.5) that $\mathrm{diag}(x) \otimes \mathrm{diag}(y) = \mathrm{diag}(xy^\top)$.

From

$$(A \times B)\mathbf{1}_V\mathbf{1}_H^\top = (A\mathbf{1}_V)\mathbf{1}_H^\top + \mathbf{1}_V(B\mathbf{1}_H)^\top$$

we then obtain

$$\mathrm{diag}\big((A \times B)\mathbf{1}_V\mathbf{1}_H^\top\big) = \mathrm{diag}(A\mathbf{1}_V) \otimes I_H + I_V \otimes \mathrm{diag}(B\mathbf{1}_H).$$

Thus,

$$\begin{aligned} L(A \times B) &= \mathrm{diag}(A\mathbf{1}_V) \otimes I_H + I_V \otimes \mathrm{diag}(B\mathbf{1}_H) - A \otimes I_H - I_V \otimes B \\ &= L(A) \otimes I_H + I_V \otimes L(B) = L(A) \times L(B), \end{aligned}$$

thereby proving the thesis. □

The spectrum of a product of matrices is determined by the spectra of the factors via a simple relation.

Proposition 1.15 (Spectrum of product matrices) *If A and B have eigenvalues λ and μ with corresponding eigenvectors $x \in \mathbb{R}^V$ and $y \in \mathbb{R}^H$ respectively, then $A \times B$ has eigenvalue $\lambda + \mu$ with eigenvector $xy^\top \in \mathbb{R}^{V \times H}$.*

Proof From the definition of Cartesian product, we observe

$$(A \times B)xy^\top = (Ax)y^\top + x(By)^\top = (\lambda + \mu)xy^\top,$$

which gives the thesis. □

The spectral properties of several families of graphs can be studied using the product operation defined above.

Example 1.6 (*Hypercube graph*) The hypercube H_n, defined in Example 1.3, is the Cartesian product of n factors K_2. The Laplace spectrum of K_2 is $\{0, 2\}$, and hence the Laplace spectrum of H_n consists of the numbers $2i$ with multiplicity $\binom{n}{i}$, for $i \in \{0, \dots, n\}$.

Other examples are products of cycle graphs: The product of two cycles represents a square lattice on a two-dimensional torus.

Example 1.7 (*Bidimensional torus grid*) We know from Example 1.5 that the graph C_n has Laplace spectrum

$$\lambda_k = 2 - 2\cos\left(\frac{2\pi}{n}k\right) \qquad k \in \{0, \dots, n-1\}.$$

Then, the product graph $C_n \times C_m$ has Laplace spectrum

$$4 - 2\cos\left(\frac{2\pi}{m}h\right) - 2\cos\left(\frac{2\pi}{n}k\right) \qquad h \in \{0, \dots, m-1\},\ k \in \{0, \dots, n-1\}.$$

The extension to k-dimensional grids is now natural.

Example 1.8 (*k-dimensional torus grid*) Since the graph C_m has Laplace spectrum $\lambda_k = 2 - 2\cos\left(\frac{2\pi}{m}k\right)$, $k \in \{0, \dots, m-1\}$, the Laplace spectrum of the product graph $C_m^k = \underbrace{C_m \times \dots \times C_m}_{k \text{ times}}$ is $2k - 2\sum_{i=1}^k \cos\left(\frac{2\pi}{m}h_i\right)$, $h_i \in \{0, \dots, m-1\}$.

The reader may compute as an exercise the spectra of other multi-dimensional graphs, e.g., L_n^k (using Example 1.27) and $\vec{C}_n^k$ (using Example 1.5).

1.5.3 Cayley Graphs

A generalization of cycle graphs and toroidal grids is provided by the family of Abelian Cayley graphs, which are graphs whose set of nodes is an Abelian group and the set of edges is stable by translation operations. The formal definition is proposed below.

Definition 1.2 (*Abelian Cayley matrices and graphs*) Let Γ be an Abelian group (we use the additive notation) and let S be a subset of Γ. Then, the *Γ-Cayley graph* generated by S in Γ is the graph $\mathscr{G}(\Gamma, S)$ having Γ as node set and

$$E = \{(g, h) \in \Gamma \times \Gamma \,:\, h - g \in S\}$$

as edge set. In words, two nodes—i.e., two group elements—are neighbors if their difference is in S. When it causes no confusion, we shall simply refer to Cayley graphs without explicitly mentioning Γ. As well, a notion of Cayley matrix can be defined. Given a group Γ and a generating row vector $\pi \in \mathbb{R}^{\Gamma}$, we shall define the *$\Gamma$-Cayley matrix* generated by π as the matrix $\mathrm{cayl}(\pi) \in \mathbb{R}^{\Gamma\times\Gamma}$ defined by $\mathrm{cayl}(\pi)_{gh} = \pi_{h-g}$ for all h and g in Γ. Correspondingly, for a given Cayley matrix M, we shall denote by π^M the generating vector of the Cayley matrix M which is simply the row of M labeled by $g = 0$.

Clearly, the adjacency and Laplacian matrices of Γ-Cayley graphs are Γ-Cayley matrices. Conversely, if P is a Γ-Cayley matrix generated by π, then $\mathscr{G}_P$ is a Γ-Cayley graph with $S = \{h \in \Gamma \,:\, \pi_h \neq 0\}$.

Abelian Cayley graphs encompass several important examples.

Example 1.9 Let $\mathbb{Z}_n$ denote the cyclic group of integers modulo n.

(i) The *complete* graph on n nodes is $\mathscr{G}(\mathbb{Z}_n, \mathbb{Z}_n \setminus \{0\})$;
(ii) The *circulant* graphs (resp. matrices) are Abelian Cayley graphs (resp. matrices) on the group $\mathbb{Z}_n$: We have that $\mathrm{cayl}(\pi) = \mathrm{circ}(\pi)$. For instance, the *cycle* graph C_n is the circulant graph $\mathscr{G}(\mathbb{Z}_n, \{-1, 1\})$; its adjacency matrix is $A = \mathrm{circ}([0, 1, 0, \dots, 0, 1])$ and its Laplacian is $L = \mathrm{circ}([2, -1, 0, \dots, 0, -1])$.
(iii) The *grids* on a d-dimensional torus are $\mathscr{G}(\mathbb{Z}_n^d, \{e_i, -e_i\}_{i\in\{1,\dots,d\}})$, where e_i are the elements of the canonical basis of $\mathbb{R}^d$.
(iv) Keeping the same notation, the *d-dimensional hypercube* is $\mathscr{G}(\mathbb{Z}_2^d, \{e_i\}_{i\in\{1,\dots,d\}})$.

The algebraic structure of Cayley graphs and matrices implies strong properties. The next two results list some basic properties which can be proven as exercises.

Proposition 1.16 *Assume Γ is an Abelian group and $S \subset \Gamma$. Then, the following statements hold true.*

(i) $\mathscr{G}(\Gamma, S)$ is a symmetric graph if and only if S is inverse-closed, and is strongly connected if and only if S generates the group Γ.
(ii) $\mathscr{G}(\Gamma, S)$ is topologically balanced and $|S|$-regular.

Proposition 1.17 *If M and M' are Γ Cayley matrices, then*

(i) their sum is $M + M'$ is Cayley and $\pi^{M+M'} = \pi^M + \pi^{M'}$;
(ii) M and M' commute and their product is Cayley. Namely, $\pi^{MM'} = \pi^M * \pi^{M'}$, where $*$ denotes convolution between vectors: $(v * v')_i = \sum_j v_j v'_{i-j}$. Moreover, $\pi^M * \pi^{M'} = M\pi^{M'}$.

The spectrum of a Cayley matrix can be computed by a discrete Fourier transform of its generating vector.

Proposition 1.18 (Spectrum of Cayley matrices) *Let $\Gamma = \mathbb{Z}_{n_1} \oplus \cdots \oplus \mathbb{Z}_{n_d}$, so that $\sum_{\ell=1}^{d} n_\ell = N$. Let M be Γ-Cayley and $\pi \in \mathbb{R}^\Gamma$ be its generating vector. Then, the spectral structure of M can be described as follows:*

(i) the eigenvalues of M are

$$\lambda_h = \sum_{k\in\Gamma} \pi_k \exp\left(-i\, 2\pi \sum_{\ell=1}^{d} \frac{k_\ell h_\ell}{n_\ell}\right) \quad h \in \Gamma;$$

(ii) a corresponding orthogonal basis of eigenvectors $\chi^{(h)} \in \mathbb{R}^\Gamma$ is given by

$$\chi^{(h)}(k) = \exp\left(i\, 2\pi \sum_{\ell=1}^{d} \frac{k_\ell h_\ell}{n_\ell}\right) \tag{1.9}$$

(iii) the matrix M can be written as

$$M = \sum_{h\in\Gamma} \lambda_h N^{-1} \chi^{(h)} \chi^{(h)*}.$$

Proof To provide this proof, we first need to briefly review part of the theory of Fourier transforms on discrete groups: We refer to [25] for a comprehensive introduction. Let $\mathbb{C}^*$ be the multiplicative group of the nonzero complex numbers. A *character* on Γ is a group homomorphism $\chi : \Gamma \to \mathbb{C}^*$, namely a function from Γ to $\mathbb{C}^*$ such that $\chi(g + h) = \chi(g)\chi(h)$ for all $g, h \in \Gamma$. We can interpret a character as a linear function $\chi : \Gamma \to \mathbb{C}^\Gamma$, i.e., as an N-dimensional vector of complex numbers. Since we have that $\chi(g)^N = \chi(Ng) = \chi(0) = 1$ for every $g \in \Gamma$, it follows that χ takes values on the Nth-roots of unity. The character $\chi_0(g) = 1$ for every $g \in \Gamma$ is called the trivial character (notice that χ_0 corresponds to $\mathbf{1}$). The set of all characters of the group Γ forms an Abelian group with respect to the entrywise multiplication. It is called the character group and denoted by $\hat{\Gamma}$. The trivial character is clearly the zero of $\hat{\Gamma}$. Moreover, $\hat{\Gamma}$ is isomorphic to Γ, and its cardinality is N. If we consider the vector space $\mathbb{C}^\Gamma$ of all functions from Γ to $\mathbb{C}$ with the canonical Hermitian form $\langle f_1, f_2\rangle = \sum_{g\in\Gamma} f_1(g) f_2(g)$, it can be proved that the set $\{N^{-1/2}\chi : \chi \in \hat{\Gamma}\}$ is an orthonormal basis of $\mathbb{C}^\Gamma$. Then, it is possible to define the Fourier transform of a function $f : \Gamma \to \mathbb{C}$ as

$$\hat{f} : \hat{\Gamma} \to \mathbb{C} \qquad \hat{f}(\chi) = \sum_{g \in \Gamma} \chi(-g) f(g).$$

After this review, consider the Cayley matrix $M \in \mathbb{R}^{\Gamma \times \Gamma}$ generated by the vector $\pi \in \mathbb{R}^{\Gamma}$. The matrix M can be interpreted as a map from $\mathbb{C}^{\Gamma}$ to itself, and the spectral structure of M can be described as follows. Each character χ is an eigenfunction of M with eigenvalue $\hat{\pi}(\chi)$, because

$$\begin{aligned}(M\chi)(g) &= \sum_{h \in \Gamma} M_{gh} \chi(h) \\ &= \sum_{h \in \Gamma} \pi_{g-h} \chi(h) \\ &= \sum_{h \in \Gamma} \pi_{g-h} \chi(g) \chi(h-g) \\ &= \sum_{\ell \in \Gamma} \pi_{\ell} \chi(-\ell) \chi(g) \\ &= \hat{\pi}(\chi) \chi(g).\end{aligned}$$

Since the characters form an orthonormal basis, it follows that M is diagonalizable and its spectrum is given by $\{\hat{\pi}(\chi) \, : \, \chi \in \hat{\Gamma}\}$. Furthermore, the matrix can be rewritten as

$$M = \sum_{\chi \in \hat{\Gamma}} \hat{\pi}(\chi) N^{-1} \chi \chi^*,$$

where $\frac{1}{N} \chi \chi^*$ is a linear function from $\mathbb{C}^{\Gamma}$ to itself, projecting $\mathbb{C}^{\Gamma}$ onto the eigenspace generated by χ. Finally, a straightforward verification shows that characters of Γ are explicitly given by the $\chi^{(h)}$'s defined in (1.9). This completes the proof. □

1.5.4 De Bruijn Graphs

We now present a remarkable class of graphs based on a combinatorial construction. A De Bruijn graph on k symbols of dimension h is defined as follows. The node set is the set all strings of length h of k given symbols, and there is an edge from u to v if v can be obtained from u by shifting all symbols by one place to the left and adding a new symbol at the rightmost place. More formally, $V = \{0, \dots, k-1\}^h$ and

$$E = \{(u, v) \in V \times V \, : \, v_{i-1} = u_i \quad \text{for all } i \in \{1, \dots, h\}\}.$$

See Fig. 1.10 for an example. An equivalent definition is given in Exercise 1.8.

De Bruijn graphs have notable properties: Here, we limit ourselves to observe the following facts, whose simple proofs are left to the reader (Exercise 1.32).

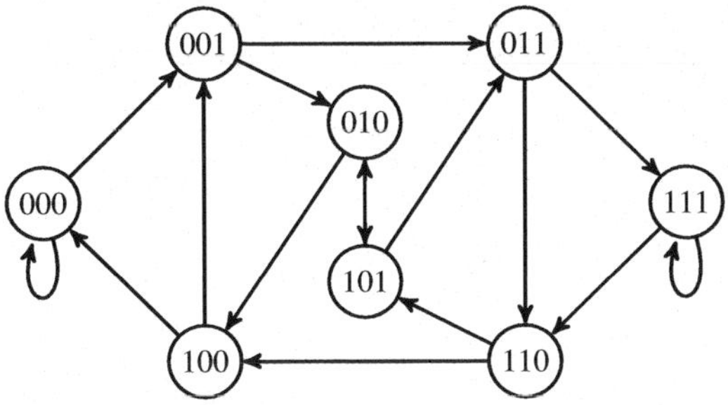

$$A = \begin{bmatrix} 1 & 1 & 0 & 0 & 0 & 0 & 0 & 0 \\ 0 & 0 & 1 & 1 & 0 & 0 & 0 & 0 \\ 0 & 0 & 0 & 0 & 1 & 1 & 0 & 0 \\ 0 & 0 & 0 & 0 & 0 & 0 & 1 & 1 \\ 1 & 1 & 0 & 0 & 0 & 0 & 0 & 0 \\ 0 & 0 & 1 & 1 & 0 & 0 & 0 & 0 \\ 0 & 0 & 0 & 0 & 1 & 1 & 0 & 0 \\ 0 & 0 & 0 & 0 & 0 & 0 & 1 & 1 \end{bmatrix}$$

Fig. 1.10 De Bruijn graph of dimension three on two symbols and its adjacency matrix

(i) If $h = 1$, then the De Bruijn graph is the complete graph.
(ii) Each vertex has exactly k incoming and k outgoing edges (counting self-loops), that is, $\mathbf{1}^* A = k\mathbf{1}^*$ and $A\mathbf{1} = k\mathbf{1}$.
(iii) The adjacency matrix A of a De Bruijn graph is such that $A^h = \mathbf{1}\mathbf{1}^*$.
(iv) The Laplacian eigenvalues of a De Bruijn graph are 0 and k, and k has multiplicity $n - 1$. In particular, the smallest nonzero Laplacian eigenvalue is equal to k.

Exercises

Exercises are divided into three groups, respectively devoted to basic graph theory (that is, the first three sections), algebraic graph theory (Sect. 1.4) and to significant examples of graphs (Sect. 1.5).

Basic Graph Theory

Exercise 1.1 (*Handshaking Lemma*) Let $G = (V, E)$ be a symmetric graph. Show that the sum of the degrees of all nodes is even.

Exercise 1.2 (*Strongly connected components*) Let there be a graph $G = (V, E)$. Verify that the relation between nodes "u and v communicate" is an *equivalence relation* in V, namely the relation is reflexive, symmetric, and transitive. Show that the strongly connected components of G are the equivalence classes of this relation.

Exercise 1.3 (*Periodicity and connectivity*) Let $G = (V, E)$ be a strongly connected graph.

(i) Prove that all nodes in G have the same period.
(ii) Prove that if at least one edge is symmetric (i.e., $\{(u, v), (v, u)\} \subset E$), then the period is either 1 or 2.
(iii) What can we argue about the period if $\vec{C}_3$ is a subgraph of G? And if $\vec{C}_4 \subset G$?

(iv) Prove Corollary 1.1.

Exercise 1.4 (*Trees have leaves*) In a symmetric graph, a vertex of degree one is said to be a *leaf*. Show that every tree has at least two leaves.

Exercise 1.5 (*Trees and number of edges*) Let $G = (V, E)$ be a symmetric graph and denote $v = |V|$ and $e = |E|/2$. Then, the following statements are true.

(i) If G is connected, then $e \geq v - 1$.
(ii) If G is connected and has at least one cycle, then $e \geq v$.
(iii) If G is a tree, then $e = v - 1$. Vice versa, if G is connected and $e = v - 1$, then G is a tree.
(iv) If G is a forest with k connected components, then $e = v - k$.
(v) If G is a connected unicycle, then $e = v$. Vice versa, if G is connected and $e = v$, then G is a unicycle.

Exercise 1.6 (*Globally reachable node*) Let $G = (V, E)$ be a graph of order at least two. Given a subset of nodes $U \subset V$, we say that v is an out-neighbor of U if there exists $(u, v) \in E$ with $u \in U$. Prove the following fact. The graph G has no globally reachable node if and only if there exist two disjoint nonempty subsets of nodes $U, W \subset V$ such that neither U nor W has an out-neighbor.

Exercise 1.7 (*Rooted and spanning trees*) A *rooted tree* is a circuit-free graph with the following property: There exists a vertex, called the *root*, such that any other vertex of the graph can be reached by one and only one directed path starting at the root. A *spanning tree* of a given graph is a spanning subgraph that is a rooted tree.

(i) A graph contains a spanning tree if and only if the reverse graph contains a globally reachable vertex.
(ii) If a graph is strongly connected, then it contains a globally reachable vertex and a spanning tree.

Exercise 1.8 (*Eulerian paths and circuits*) An Eulerian path is a path that visits all the graph edges exactly once. An Eulerian circuit is an Eulerian path which starts and ends at the same node. Show that the following properties hold for a graph G.

(i) If G has an Eulerian circuit, then G is topologically balanced.
(ii) If G is weakly connected and has an Eulerian circuit, then G is strongly connected.
(iii) G has an Eulerian circuit if and only if G is topologically balanced, and all of its vertices with nonzero degree belong to the same strongly connected component.
(iv) Let G be a weakly connected graph. Then, G is topologically balanced if and only if G has an Eulerian circuit.
(v) A weakly connected graph $G = (V, E)$ has an Eulerian path if and only if

   (a) at most one vertex $v \in V$ has $d_v^{\text{out}} - d_v^{\text{in}} = 1$;
   (b) at most one vertex $u \in V$ has $d_u^{\text{in}} - d_u^{\text{out}} = 1$; and
   (c) every other vertex has equal in-degree and out-degree.

Algebraic and Spectral Graph Theory

Exercise 1.9 (*Bipartite graph and adjacency matrices*) Observe that the adjacency matrix of bipartite graph $G = (V, E)$ can be written, up to a vertex permutation, in a block form

$$A_G = \begin{bmatrix} 0 & B \\ C & 0 \end{bmatrix}$$

where $B \in \mathbb{R}^{V_1 \times V_2}$ and $C \in \mathbb{R}^{V_2 \times V_1}$, such that $V_1 \cap V_2 = \emptyset$, $V_1 \cup V_2 = V$.

(i) Compute $A_G^2, A_G^3, \ldots$
(ii) Prove that there is no path in G of odd length.

Exercise 1.10 (*Weight balance and connectivity*) Let G be a weighted graph. If G is weight-balanced and contains a globally reachable node, then G is strongly connected.

Exercise 1.11 (*Distance on a weighted graph*) Let $G = (V, E, A)$ be a weighted graph, and define the length of a path as the sum of the weights associated to the edges insisting on the path. Then, assuming that G is connected, define the distance function $\mathrm{dst} : V \times V \to \mathbb{R}_{\geq 0}$ as in (1.2). Observe that these definitions naturally extend the notions of path length and distance between nodes to weighted graphs. Prove that, if G is symmetric, then dst is a *metric* on V, that is

(i) $\mathrm{dst}(u, v) \geq 0$ and $\mathrm{dst}(u, v) = 0$ if and only if $u = v$ (positive definiteness);
(ii) $\mathrm{dst}(u, v) = \mathrm{dst}(v, u)$ for every $u, v \in V$ (symmetry);
(iii) $\mathrm{dst}(u, v) \leq \mathrm{dst}(u, w) + \mathrm{dst}(w, v)$ for every $u, v, w \in V$ (triangle inequality).

Exercise 1.12 (*Directed incidence matrix*) Let $G = (V, E, A)$ be a weighted graph and $L = L(A)$ the corresponding Laplacian matrix. Define the *directed incidence matrix* $\Phi \in \mathbb{R}^{E \times V}$ by

$$\Phi_{(u,v),w} = \begin{cases} 1 & \text{if } w = u \\ -1 & \text{if } w = v \\ 0 & \text{otherwise} \end{cases}$$

so that for each row, corresponding to an edge, there is a 1 corresponding to the tail and a -1 corresponding to the head of the edge. Then, show that

$$\Phi^* W \Phi = L + L^*,$$

where $W \in \mathbb{R}^{E \times E}$ is a diagonal matrix arranging all the weights of the edges, such that $W_{(u,v),(u,v)} = A_{uv}$.

Exercise 1.13 (*The 4-wheel*) Let $G = (V, E)$ be symmetric and $|V| = 5$.

(i) Prove or disprove the existence of a graph on V such that

  (a) all vertices have degree 3;
  (b) four vertices have degree 3 and one vertex has degree 4.

(ii) When such graph exists, may it be bipartite? May it contain a Eulerian circuit?
(iii) Write the adjacency matrix of such a graph, and compute its eigenvalues.

Exercise 1.14 (*Spectra of regular graphs*) Let G be a graph and assume G is out-regular with out-degree d. Then, λ is an adjacency eigenvalue if and only if $d - \lambda$ is a Laplacian eigenvalue.

Exercise 1.15 (*Complement graph*) Given a graph $G = (V, E)$, let the *complement* graph $\bar{G}$ be the graph having node set V of cardinality n and edge set $\bar{E} = \{(u, v) \in V \times V : u \neq v \text{ and } (u, v) \notin E\}$.

(i) Show that the Laplacian matrices of G and $\bar{G}$ are such that

$$L(G) + L(\bar{G}) = nI - \mathbf{1}\mathbf{1}^*.$$

(ii) For a given graph H, let $\lambda_j(H)$ denote the j-th smallest eigenvalue of $L(H)$. Conclude from (i) that for $2 \leq j \leq n$, it holds $\lambda_j(\bar{G}) = n - \lambda_{n+2-j}(G)$.
(iii) Prove that G and $\bar{G}$ cannot both be disconnected.

Exercise 1.16 (*Edge addition*) Let $G = (V, E)$ be a symmetric graph and $0 = \lambda_1(G) \leq \lambda_2(G) \leq \cdots \leq \lambda_n(G)$ its Laplacian spectrum. Let $G \cup \{e\}$ denote the addition of a pair of edges $\{(u, v), (v, u)\}$ to the graph G. Then,

$$\lambda_2(G) \leq \lambda_2(G \cup \{e\}) \leq \lambda_2(G) + 2.$$

Exercise 1.17 (*Bounds on Laplacian eigenvalues*) Let $G = (V, E)$ be a symmetric graph and λ an eigenvalue of its Laplacian. Show that $\lambda \leq 2d_{\max}$ and that $\lambda \leq n$, where $d_{\max}$ is the largest degree and n the order of G. Find examples where the corresponding equalities hold.

Exercise 1.18 (*Algebraic connectivity and vertex connectivity*) For a connected symmetric graph G which is not complete, we define the *vertex connectivity* of G as

$$\kappa(G) = \min\{k \in \mathbb{N} : G \text{ can be disconnected by removing } k \text{ nodes}\}.$$

Then, $\lambda_2 \leq \kappa(G)$, where λ_2 denotes the smallest nonzero eigenvalue of L_G. For this reason, λ_2 also takes the name of *algebraic connectivity* of the graph.

Exercise 1.19 (*Information from spectrum*) Let $G = (V, E)$ be symmetric and regular, and assume that the spectrum of the adjacency matrix of G is

$$\{-3, -3, -1, -1, -1, -1, 1, 1, 0, 2, 2, 4\}.$$

(i) Compute the cardinality of V.
(ii) Compute the degree of G.
(iii) Is G connected?

(iv) Is G bipartite?
(v) Estimate the diameter of G.

Exercise 1.20 (*Bipartite graphs*) Let M be adapted to a bipartite graph. Show that if λ is an eigenvalue of M, then also $-\lambda$ is an eigenvalue of M.

Exercise 1.21 (*Properties of Laplacians*) Let $G = (V, E, A)$ be a weighted graph of order n and L be the (weighted) Laplacian of G. The following statements hold true.

(i) All eigenvalues of L have nonnegative real part.
(ii) The following properties are equivalent:

  (a) G is weight-balanced;
  (b) $\mathbf{1}^* L = 0$;
  (c) for all $x \in \mathbb{R}^V$, it holds $x^*(L + L^*)x = \sum_{u,v} A_{uv}(x_v - x_u)^2$;
  (d) $L + L^*$ is positive semidefinite.

(iii) G is weakly connected if and only if $\ker(L + L^*) = \text{span}\{\mathbf{1}\}$.

Exercise 1.22 (*Diameter and Laplacian* [1, 20]) Given a connected symmetric graph G of order n, let λ_2 be its smallest nonzero Laplacian eigenvalue. Then, $\lambda_2 \geq \frac{4}{n(n-1)}$ and

$$\frac{4}{n\lambda_2} \leq \text{diam}(G) \leq 2\left\lceil \sqrt{\frac{2d_{\max}}{\lambda_2}} \log_2 n \right\rceil.$$

Exercise 1.23 (*Diameter and normalized Laplacian*) Let $G = (V, E)$ be a symmetric graph of order n, A its adjacency matrix, and D its (diagonal) degree matrix. Let also $L = D - A$ be the Laplacian matrix and $M = I - D^{-1/2} A D^{-1/2}$ the *normalized Laplacian* (according to [10]). We denote by $0 = \mu_1 \leq \cdots \leq \mu_n \leq 2$ the eigenvalues of M. Then,

$$\frac{1}{\mu_2 |E|} \leq \text{diam}(G) \leq \frac{\log n}{\log \frac{\mu_n + \mu_2}{\mu_n - \mu_2}}.$$

Exercise 1.24 (*Paths of a certain length*) Let G be a graph of order n and assume there exists $h \in \mathbb{N}$ such that there is exactly one path of length h connecting any pair of nodes. Prove that there exists k such that

(i) $n = k^h$;
(ii) each vertex of G has exactly k in-neighbors and out-neighbors;
(iii) k is the only nonzero adjacency eigenvalue of G and has multiplicity one.

Example Families of Graphs

Exercise 1.25 (*Star graph*) The *star graph* S_n is a symmetric graph of order $n + 1$ such that one node, called the center, is the only neighbor to all the other n nodes; see

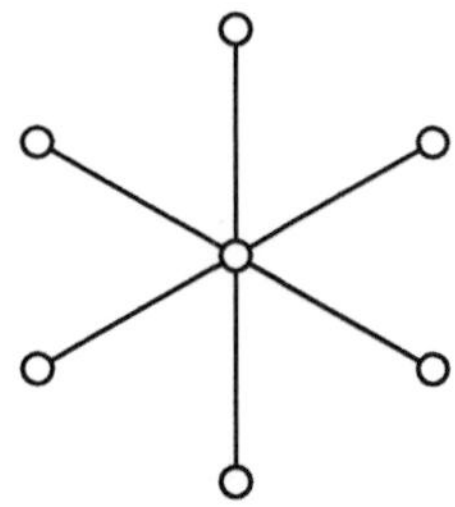

Fig. 1.11 The star graph S_6

Fig. 1.11. Observe that S_n is a tree and is isomorphic to $K_{1,n}$; its diameter is two for every n. Let $A_n = \begin{bmatrix} 0 & \mathbf{1}_n^* \\ \mathbf{1}_n & 0 \end{bmatrix}$ denote its adjacency matrix and L_n denote its Laplacian. Show that

(i) powers of A_n have the form

$$A_n^{2k-1} = n^{k-1} A_n, \qquad A_n^{2k} = n^{k-1} \begin{bmatrix} n & 0 \\ 0 & \mathbf{1}_n \mathbf{1}_n{}^* \end{bmatrix} \qquad \text{for all } k \in \mathbb{N};$$

(ii) the eigenvalues of L_n are the solutions of the polynomial

$$p(s) = s(s-1)^{n-1}(s-n).$$

Exercise 1.26 (*Kronecker product*) Verify the following useful properties of the Kronecker product of matrices as defined in (1.5). All matrices are assumed to be square.

(i) $AB \otimes CD = (A \otimes C)(B \otimes D)$;
(ii) $(A \otimes B)^* = (A^* \otimes B^*)$;
(iii) $(A \otimes B) \otimes C = A \otimes (B \otimes C)$, up to a permutation of indices;
(iv) $A \otimes B = B \otimes A$, up to a permutation of indices;
(v) $(A \otimes B)^{-1} = (A^{-1} \otimes B^{-1})$, provided A and B are invertible;
(vi) the eigenvalues of $A \otimes B$ are all possible products of an eigenvalue of A with an eigenvalue of B;
(vii) the eigenvectors of $A \otimes B$ are all possible Kronecker products of an eigenvector of A with an eigenvector of B;
(viii) $\operatorname{tr} A \otimes B = \operatorname{tr} A \operatorname{tr} B$;
(ix) $\det(A \otimes B) = (\det A)^n (\det B)^m$ where n and m are the dimensions, respectively, of A and B.

Exercise 1.27 (*Tridiagonal and augmented tridiagonal Toeplitz matrices* [6, Lemma 1.77 and (1.6.7)]) Let $V = \{1, \ldots, n\}$ and $a, b \in \mathbb{R}$. Consider the $n \times n$ *tridiagonal Toeplitz* matrix

$$\mathrm{Trid}_n(a,b) = \begin{bmatrix} b & a & 0 & \dots & & 0 \\ a & b & a & 0 & \cdots & 0 \\ 0 & a & b & a & \cdots & 0 \\ \vdots & & \ddots & \ddots & \ddots & \\ & \cdots & 0 & a & b & a \\ 0 & \cdots & & 0 & a & b \end{bmatrix}.$$

(i) The matrix $\mathrm{Trid}_n(a,b)$ has eigenvalues $\lambda_i = b + 2a\cos\left(\frac{i\pi}{n+1}\right)$, $i \in \{1,\dots,n\}$ and corresponding eigenvectors

$$x^{(i)} = \left[\sin\left(\frac{i\pi}{n+1}\right), \sin\left(\frac{2i\pi}{n+1}\right), \dots, \sin\left(\frac{ni\pi}{n+1}\right)\right]^*.$$

Furthermore, consider the $n \times n$ *augmented tridiagonal* Toeplitz matrix

$$\mathrm{ATrid}_n^{\pm}(a,b) := \mathrm{Trid}_n(a,b) \pm \begin{bmatrix} a & 0 & \dots & & 0 \\ 0 & 0 & \dots & & 0 \\ \vdots & & & \ddots & \ddots \\ 0 & \cdots & & 0 & a \end{bmatrix}$$

(ii) The matrix $\mathrm{ATrid}_n^{\pm}(a,b)$ has eigenvalues

$$\lambda_i = b + 2a\cos\left(\frac{i\pi}{n}\right) \qquad i \in \{1,\dots,n-1\},$$
$$\lambda_n = b \pm 2a.$$

Exercise 1.28 (*Bidimensional grid*)

(i) Use Exercise 1.27 to verify that the graph L_m has adjacency spectrum

$$2\cos\left(\frac{k\pi}{m+1}\right) \qquad k \in \{1,\dots,m\}.$$

(ii) Show that the product graph $L_m \times L_n$ has adjacency spectrum

$$2\cos\left(\frac{2\pi}{m+1}h\right) + 2\cos\left(\frac{2\pi}{n+1}k\right) \qquad h \in \{1,\dots,m\},\ k \in \{1,\dots,n\}.$$

(iii) Find the Laplacian spectrum of L_m and $L_m \times L_n$.

Exercise 1.29 (*Cayley graph that is not a product*) Consider the Γ-Cayley graph G where $\Gamma = \mathbb{Z}_N \times \mathbb{Z}_N$ and where

$$S = \{(\pm 1, 0),\ (0, \pm 1),\ \pm(1,1)\}.$$

Verify that the Laplacian eigenvalues are given by

$$6 - 2\cos\frac{2\pi h}{N} - 2\cos\frac{2\pi k}{N} - 2\cos\frac{2\pi(h+k)}{N}, \quad h, k = 0, \dots, N-1.$$

Exercise 1.30 (*Algebraic connectivity of symmetric Cayley graphs*) Let λ_2 be the algebraic connectivity (cf. Exercise 1.18) of the Cayley graph $\mathcal{G}(\Gamma, S)$. Then, a nontrivial result in [7] implies that

$$\lambda_2 \leq \frac{C}{|\Gamma|^{2/|S|}},$$

where $C > 0$ is a constant independent of Γ and S. This inequality can easily be verified in examples of Cayley graphs presented in the text. For instance, you can check that

(i) the Laplacian eigenvalues of C_n are $\{2(1 - \cos(\frac{2\pi}{n}\ell))\}_{\ell \in \mathbb{Z}_n}$, and in particular

$$\lambda_2 = 2 - 2\cos\left(\frac{2\pi}{n}\right) \leq \frac{2\pi^2}{n^2};$$

(ii) the Laplacian eigenvalues of C_m^2 are $4 - 2\cos(\frac{2\pi}{m}h) + \cos(\frac{2\pi}{m}k)$, for $h, k \in \{0, \dots, m-1\}$ and in particular

$$\lambda_2 = 4 - 4\cos\left(\frac{2\pi}{m}\right) \leq \frac{2\pi^2}{m^2};$$

(iii) the Laplacian eigenvalues of H_d are $\{2\ell\}_{\ell \in \{0,\dots,d\}}$, and in particular $\lambda_2 = 2$.

Exercise 1.31 (*De Bruijn graphs*) Show that the De Bruijn graph on k symbols of dimension h denoted as B_k^h is the graph with order $n = k^h$ such that every node $u \in \{0, \dots, n-1\}$ is connected to $ku, ku+1, ku+2, \dots, ku+k-1$ (all mod k^h).

Exercise 1.32 (*Properties of De Bruijn graphs*) By using Exercise 1.24, prove the properties of De Bruijn graphs stated in Sect. 1.5.4.

Exercise 1.33 (*Geometric graphs*) Let V be a node set and $x \in (\mathbb{R}^d)^V$. The *r-disk graph* is a symmetric graph $\mathcal{G}_{r,\mathrm{disk}}(x) = (V, E(x))$ defined by $E(x) = \{(u, v) : \|x_v - x_u\| \leq r\}$. On the other hand, define the *distance graph* as the complete graph endowed with a weight matrix $W(x)$ such that $W_{uv}(x) = \|x_u - x_v\|$. Define the *Euclidean minimum spanning tree* $\mathcal{G}_{\mathrm{EMST}}(x)$ as the spanning tree of the distance graph of minimum weight (i.e., such that the sum of the weights of its edges is minimal). Show that $\mathcal{G}_{\mathrm{EMST}}(x) \subset \mathcal{G}_{r,\mathrm{disk}}(x)$ if and only if $\mathcal{G}_{r,\mathrm{disk}}(x)$ is connected.

Bibliographical Notes

Our account of graph theory is of course far from being a complete one. Instead, we have selected definitions and facts that will be needed in the following chapters. Hence, we expect that the reader may be interested in a broader introduction, for instance the books [3, 13]. Moreover, previous books on network coordination and robotic networks do contain introductions to graph theory, which partly overlap with ours [6, 19]. Our definition of graph is often referred to as *directed* graph in the literature, as opposed to *undirected* graphs, pairs $(V, \bar{E})$ in which the elements of $\bar{E}$ are unordered pairs of nodes $\{u, v\}$. Then, an undirected graph is equivalent (in our language) to a symmetric graph in which each pair of directed edges (u, v), (v, u) is counted as one. This notion of undirected graph will be used later in Chap. 5.

Matrices adapted to graphs are fundamental in this book: For this reason, we have devoted significant attention to algebraic graph theory, which studies graphs via certain matrices associated to them, especially the adjacency and the Laplacian matrices. The study of their spectra is the goal of spectral graph theory: Several books on this topic are available [5, 10, 11]. In our treatment, we have focused on the Laplacian spectrum: The properties of the adjacency spectrum are also notable [3, Sect. VIII], but less useful to our needs (for instance, insufficient to characterize connectivity [5, Sect. 1.3.7]).

Section 1.5 has been devoted to selected examples of graphs and associated matrices. We have also introduced the Cartesian product, a useful operation to construct graphs [9, 17]. We recall here the examples that we presented, together with some useful references. *Circulant* matrices are a standard topic in applied mathematics, covered for instance in the classical book [12]. A related class of matrices is that of *Toeplitz* matrices: General Toeplitz matrices are not important to us, but an example that is useful to compute the spectrum of line graphs is presented in Exercise 1.27. More generally, Cayley graphs have a long history in abstract mathematics and have been used in control theory to describe translation-invariant systems [24]. Our interest in *Abelian Cayley* topologies is motivated both by their algebraic properties, which allow for an elegant mathematical treatment [7, 25], and by their potential for the applications. Indeed, Abelian Cayley graphs are idealized representations of communication scenarios of practical interest. In particular, they describe communication patterns that are *local*, not only in the sense of a limited number of neighbors, but also with a bound on the geometric distance among connected nodes. In Abelian Cayley graphs, this constraint is abstracted into the definition of edge set [2, 4, 8, 15, 18]. For this reason, Abelian Cayley graphs are an alternative to other models of "local" communication, such as geometric graphs. *Geometric graphs* are graphs such that each node is endowed with a location and the edge set depends on these locations (see Exercise 1.33); various types of geometric graphs are presented in [6, Chap. 2]. If a geometric graph is constructed from a position vector which is a random variable, then its properties (e.g., connectivity) can be studied statistically. Such random geometric graphs have been studied extensively [22, 23] as a modeling paradigm to describe wireless communication networks [14, 16]. It is important to mention that other random families of graphs are used to describe different kinds of real-world networks, such as social networks, broadly referred to as to *complex networks*. Such graphs typically exhibit scale-free properties, small diameter and small spectral gap, thus being very different from geometric graphs. Their description is outside the scope of this book, but many sources are available to the interested reader [21, 26].

References

1. Alon, N., Milman, V.D.: λ_1, isoperimetric inequalities for graphs, and superconcentrators. J. Comb. Theory, Ser. B **38**(1), 73–88 (1985)
2. Bamieh, B., Jovanovic, M.R., Mitra, P., Patterson, S.: Coherence in large-scale networks: dimension-dependent limitations of local feedback. IEEE Trans. Autom. Control **57**(9), 2235–2249 (2012)
3. Bollobás, B.: Modern Graph Theory. Springer (1998)
4. Boyd, S., Ghosh, A., Prabhakar, B., Shah, D.: Randomized gossip algorithms. IEEE Trans. Inf. Theory **52**(6), 2508–2530 (2006)
5. Brouwer, A.E., Haemers, W.H.: Spectra of Graphs. Springer (2012)
6. Bullo, F., Cortés, J., Martínez, S.: Distributed Control of Robotic Networks. Applied Mathematics Series. Princeton University Press (2009)
7. Carli, R., Fagnani, F., Speranzon, A., Zampieri, S.: Communication constraints in the average consensus problem. Automatica **44**(3), 671–684 (2008)
8. Carli, R., Garin, F., Zampieri, S.: Quadratic indices for the analysis of consensus algorithms. In: Information Theory and Applications Workshop, San Diego, CA, pp. 96–104, February 2009
9. Chapman, A., Nabi-Abdolyousefi, M., Mesbahi, M.: Controllability and observability of network-of-networks via Cartesian products. IEEE Trans. Autom. Control **59**(10), 2668–2679 (2014)
10. Chung, F.R.K.: Spectral graph theory. Am. Math. Soc. (1997)
11. Cvetovic, D., Doob, M., Sachs, H.: Spectra of Graphs. Academic Press (1980)
12. Davis, P.J.: Circulant Matrices. John Wiley (1979)
13. Diestel, R.: Graph Theory, vol. 173. Graduate Texts in Mathematics. Springer, 2 edn. (2000)
14. Franceschetti, M., Meester, R.: Random Networks for Communication. Cambridge University Press (2007)
15. Garin, F., Zampieri, S.: Mean square performance of consensus-based distributed estimation over regular geometric graphs. SIAM J. Control Optim. **50**(1), 306–333 (2012)
16. Gupta, P., Kumar, P.R.: The capacity of wireless networks. IEEE Trans. Inf. Theory **46**(2), 388–404 (2000)
17. Lee, H., Linh, N.T.H., Fujisaki, Y.: Algebraic connectivity of network-of-networks having a graph product structure. Syst. Control Lett. **104**, 15–20 (2017)
18. Lovisari, E., Zampieri, S.: Performance metrics in the average consensus problem: a tutorial. Annu. Rev. Control **36**(1), 26–41 (2012)
19. Mesbahi, M., Egerstedt, M.: Graph Theoretic Methods for Multiagent Networks. Applied Mathematics Series. Princeton University Press (2010)
20. Mohar, B.: The Laplacian spectrum of graphs. Graph Theory, Comb. Appl. **2**(6), 871–898 (1991)
21. Newman, M.E.J.: Networks: An Introduction. Oxford University Press (2010)
22. Penrose, M.: Random Geometric Graphs. Oxford University Press (2003)
23. Rai, S.: The spectrum of a random geometric graph is concentrated. J. Theor. Probab. **20**(2), 119–132 (2007)
24. Recht, B., D'Andrea, R.: Distributed control of systems over discrete groups. IEEE Trans. Autom. Control **49**(9), 1446–1452 (2004)
25. Terras, A.: Fourier Analysis on Finite Groups and Applications. Cambridge University Press (1999)
26. Van Mieghem, P.: Graph Spectra for Complex Networks. Cambridge University Press (2010)

Chapter 2
Averaging in Time-Invariant Networks

Abstract This chapter studies the basic averaging dynamics on a fixed network. This linear dynamics is also called "consensus" dynamics, because under suitable assumptions it brings the states associated with the nodes to converge to the same value. Section 2.1 introduces the rendezvous problem, which serves us as the main motivation to seek consensus, and states the main results of the chapter. Section 2.2 solves the consensus problem in the special case of symmetric regular graphs, while the general solution, which is based on the notion of stochastic matrix, is presented in Sect. 2.3. The subsequent sections provide further insights into the averaging dynamics, namely about its speed of convergence (Sects. 2.4 and 2.7) and its consensus value (Sect. 2.5). Meanwhile, Sect. 2.6 presents some classical examples of stochastic matrices associated with a graph, such as simple random walks and Metropolis walks. Finally, Sect. 2.8 concentrates on reversible stochastic matrices and their properties.

2.1 Rendezvous and Consensus

One of the simplest examples of coordinated control is the so-called *rendezvous problem*. Assume that units have dynamics of type $x_v(t+1) = x_v(t) + u_v(t)$ for all $t \in \mathbb{Z}_{\geq 0}$ with $x_v(t)$ and $u_v(t) \in \mathbb{R}^n$ for all $v \in V$ and that the control goal is to make all units converge their state to the same point. We can think of them as moving agents with the state representing position. This is known as the *rendezvous problem*: There are many variants of this problem, and the one we are addressing is just the basic and simplest instance. What are exactly the issues we want to analyze? Here is a brief list:

(a) Given a graph G, find out whether there exists a control scheme $u_v = g_v(x)$ adapted to G such that the state evolutions governed by the equations $x_v(t+1) = x_v(t) + g_v(x)$ all converge to the same point, namely for all initial conditions $x(0)$, there exists $x^\star \in \mathbb{R}^n$ such that

$$\lim_{t\to+\infty} x_v(t) = x^\star\,, \quad \forall v \in V. \tag{2.1}$$

© Springer International Publishing AG 2018
F. Fagnani and P. Frasca, *Introduction to Averaging Dynamics over Networks*, Lecture Notes in Control and Information Sciences 472,
https://doi.org/10.1007/978-3-319-68022-4_2

(b) In case when (a) has a positive answer, we would like to find effective ways for producing the control scheme. Indeed, in general, there will be many possible control schemes and the choice can be dictated to optimize certain performance indices:

(b1) the velocity of convergence to the rendezvous point;
(b2) the displacement of $x^\star$ from the initial condition.

Both indices will be defined precisely later on.

Notice that without further assumptions, the problem as stated in (a) is always solvable and with no communication among units. It is sufficient to put $u_v = -x_v$, and we will have that $x_v(t) = 0$ for all v and for all $t \geq 1$: In control theory, this is known as a "deadbeat control." The reason why this is not a feasible solution is the following. This solution implicitly requires that units have already agreed to make 0 their rendezvous point, and in other terms, they have already coordinated off-line. This prior coordination is something we want to avoid; moreover, the origin may be far off from their initial condition and thus an unreasonable choice (in general not optimizing (b2)). We make the following extra assumption on the rendezvous point $x^\star$ which automatically drops out the deadbeat control scheme above: We require that, translating all initial conditions $x_v(0) \to x_v(0) + b$ with the same vector, also the rendezvous point translates the same way $x^\star \to x^\star + b$. We will refer to this as to the *translation invariance requirement.*

As it is customary in control theory, it is natural to seek, *in primis*, a linear solution to this problem, namely to consider controllers of type

$$u_v(t) = \sum_{w \in V} K_{vw} x_w(t) \tag{2.2}$$

where $K \in \mathbb{R}^{V \times V}$ is a gain matrix. Coupling with the unit dynamics, we thus obtain

$$x_v(t+1) = \sum_{w \in V} P_{vw} x_w(t) \tag{2.3}$$

where $P = I + K$.

This type of models (2.3) has applications much broader than just in the rendezvous problem for mobile agents. Instead of a position, the state $x_v(t)$ can as well be interpreted as an estimation or as an opinion on some fact possessed by unit v at time t and the common convergence to the same value is a phenomenon known as *consensus*. Later on, we will provide more details on such possible applicative contexts.

Notice that the dimension of the state does not play any particular role in the dynamics (2.3) as all components of the state vectors $x_v(t)$ evolve separately all with the same dynamics given by the matrix P. For this reason, from now on, we will assume that the state $x_v(t)$ of each unit is one-dimensional, namely a *scalar*. In this setting, (2.3) can be rewritten in more compact form simply as

$$x(t+1) = Px(t) \tag{2.4}$$

so that $x(t) = P^t x(0)$. The translation invariance, in this context, amounts to require that $P^t \mathbf{1} \to \mathbf{1}$ for $t \to +\infty$. Since $P^{t+1}\mathbf{1} = PP^t\mathbf{1}$ then converges both to $\mathbf{1}$ and to $P\mathbf{1}$, the translation invariance is also equivalent to require $P\mathbf{1} = \mathbf{1}$ (each row of P sums to 1).

Notice moreover that the feedback law (2.2) is adapted to G if K (or equivalently P) is adapted to G. Therefore, in order to exhibit a solution to the rendezvous problem with translation invariance, it is sufficient to exhibit $P \in \mathbb{R}^{V \times V}$ adapted to G such that $P\mathbf{1} = \mathbf{1}$. The following result, which will be proven in the next sections, is an elegant and simple solution.

Theorem 2.1 (Consensus) *Suppose G has a globally reachable vertex $v^\star$. Then the rendezvous problem with the translation-invariant requirement is solvable over G. A possible solution is given by any matrix $P \in \mathbb{R}^{V \times V}$ satisfying the following properties:*

(Pa) $P_{vw} \geq 0$ *for every* $v, w \in V$;
(Pb) $P\mathbf{1} = \mathbf{1}$;
(Pc) *For every* $v \neq w$, $P_{vw} > 0 \Leftrightarrow (v, w) \in E$;
(Pd) $P_{v^\star v^\star} > 0$.

It turns out that matrices as P sharing (Pa) and (Pb) have very special properties: They are called *stochastic* and appear in many different contexts, one of these being Markov chain theory. Property (Pc) says that $\mathcal{G}_P$ and G can only possibly differ in their self-loops.

There is an additional nice property of these systems. Being P stochastic, its Laplacian is $L(P) = I - P$. Consequently, we may write (2.4) as $x(t+1) = x(t) - L(P)x(t)$, which becomes $x_v(t+1) = x_v(t) + \sum_w P_{vw}(x_w(t) - x_v(t))$ componentwise. We observe that this expression only involves the state of v and differences between the states of v and of its neighbors w. Then, there is no need for the nodes to exchange information in an absolute reference frame, but only relative information suffices.

Before presenting the key results for stochastic matrices and proving Theorem 2.1 and some generalizations, we will work out a special case, which explains how matrices like P above come naturally into the picture.

2.2 Averaging on Symmetric Regular Graphs

Notice that if the underlying graph was the complete one, the rendezvous problem would have a very simple solution: It would be enough for all the units to compute the barycenter $\overline{x}(t) := N^{-1} \sum_{v \in V} x_v(t)$ (where $N := |V|$) and implement the control law $u_v(t) = \overline{x}(t) - x_v(t)$ which would yield $x_v(t) = \overline{x}(t)$ for all v and all positive t. This law implies that at time $t = 1$, all units have already reached consensus exactly in the barycenter of the initial state $\overline{x}(0)$. It is then immediate to see that $x_v(t) = \overline{x}(0)$ for every $t \geq 1$. The type of matrix P we obtain in this case is $P = N^{-1}\mathbf{1}\mathbf{1}^*$, a very special stochastic matrix with all elements equal to $1/N$.

This solution is not admissible for a general graph, but its main idea can be adapted. Indeed, it is sufficient to replace the barycenter $\overline{x}(t)$ with a local version of it, namely for each unit to use a local barycenter based on the units to which it is connected through the graph. Precisely, given a graph $G = (V, E)$, each unit $v \in V$ computes at time t

$$\overline{x}_v(t) := \frac{1}{d_v^{\text{out}}} \sum_{w \in V} (A_G)_{vw} x_w(t)$$

and implements the dynamics $x_v(t+1) = x_v(t) + \tau(\overline{x}_v(t) - x_v(t))$. The parameter $\tau > 0$ indicates the velocity at which unit v is following the local barycenter and will play a crucial role in the rest of this section. In compact matrix form, we obtain that $x(t+1) = Px(t)$ where

$$P = I + \tau(D_G^{-1} A_G - I) = I - \tau D_G^{-1} L(G). \tag{2.5}$$

It is easy to see that $\tau \in (0, 1]$ guarantees that P is a stochastic matrix adapted to the graph G.

Let us now analyze the special case when G is *symmetric and d-regular*. In this case, $P = I - \tau d^{-1} L(G)$ is also symmetric. Assuming that $0 = \lambda_1 \leq \lambda_2 \leq \cdots \leq \lambda_N$ are the eigenvalues of the Laplacian $L(G)$, we obtain that the eigenvalues of P are simply given by $\mu_i = 1 - \tau d^{-1}\lambda_i$ (with $\mu_1 = 1$). Moreover, the two matrices $L(G)$ and P share the same orthonormal basis of eigenvectors ξ_i's (with $\xi_1 = N^{-1/2}\mathbf{1}$). We can thus write the usual orthonormal decomposition of P

$$P = \sum_{i=1}^{N} \mu_i \xi_i \xi_i^* = N^{-1}\mathbf{1}\mathbf{1}^* + \sum_{i=2}^{N} \mu_i \xi_i \xi_i^*$$

which yields, by orthonormality, $P^t = N^{-1}\mathbf{1}\mathbf{1}^* + \sum_{i=2}^{N} \mu_i^t \xi_i \xi_i^*$. The evolution of the state configuration is thus

$$P^t x(0) = \bar{x}(0)\mathbf{1} + \sum_{i=2}^{N} \mu_i^t \xi_i \xi_i^* x(0)$$

Notice now that

$$||P^t x(0) - \bar{x}(0)\mathbf{1}||^2 = \left|\left|\sum_{i=2}^{N} \mu_i^t \xi_i \xi_i^* x(0)\right|\right|^2 = \sum_{i=2}^{N} |\mu_i|^{2t} |\xi_i^* x(0)|^2$$

Since $1 = \mu_1 \geq \mu_2 \geq \cdots \geq \mu_N$, putting $\rho_2 := \max\{|\mu_2|, |\mu_N|\}$, we obtain

$$||P^t x(0) - \bar{x}(0)\mathbf{1}||^2 \leq \rho_2^{2t} \sum_{i=2}^{N} |\xi_i^* x(0)|^2 \leq \rho_2^{2t} ||x(0)||^2,$$

which can be rewritten as

$$||P^t x(0) - \bar{x}(0)\mathbf{1}|| \leq \rho_2^t ||(I - N^{-1}\mathbf{1}\mathbf{1}^\top)x(0)||. \tag{2.6}$$

This bound shows that if $\rho_2 < 1$, then $x(t) = P^t x(0) \to \bar{x}(0)\mathbf{1}$ for $t \to +\infty$, namely all states converge to a consensus point, which turns out to be again the barycenter of the initial state conditions $\bar{x}(0)$. Moreover, (2.6) actually shows that ρ_2 dictates the speed of convergence of the dynamics toward consensus. Under which conditions can we guarantee that $\rho_2 < 1$? Because of the way ρ_2 is defined, we must have $|\mu_2|, |\mu_N| < 1$. If G is not connected, we know that $\lambda_2 = 0$ and, consequently, $\mu_2 = 1$: Indeed, in this case, it is clear that consensus can not be reached in general since the network is composed of completely separated components. Instead, if G is connected, then $\lambda_2 > 0$ and, consequently, $1 > \mu_2 \geq \mu_N$. Hence, the only extra condition that needs to be satisfied is $\mu_N > -1$, namely $1 - \tau d^{-1}\lambda_N > -1$. This is equivalent to $\tau < \frac{2d}{\lambda_N}$. Considering that (see Exercise 1.17) $\lambda_N \leq 2d$, a sufficient condition which guarantees consensus is $\tau < 1$. We can summarize the above discussion in the following result.

Proposition 2.1 (Consensus on symmetric regular graphs) *Let G be a symmetric, d-regular, and connected graph. Then, the dynamics* (2.4)–(2.5), *with* $\tau \in (0, 1)$, *guarantees convergence to consensus, where the consensus point is the barycenter of the initial state and convergence happens at an exponential rate given by* ρ_2.

In the next section, we will present a number of general results on stochastic matrices and we will be able to generalize this result to more general graphs, by dropping the assumptions of symmetry, regularity, and—to a certain extent—connectivity of the underlying graph.

2.3 Stochastic Matrices and Averaging

In general, a matrix $P \in \mathbb{R}^{\mathcal{V}\times\mathcal{V}}$ such that $P_{vw} \geq 0$ for every $v, w \in \mathcal{V}$ is called a *nonnegative matrix*. A nonnegative matrix $P \in \mathbb{R}^{\mathcal{V}\times\mathcal{V}}$ satisfying the row sum condition $P\mathbf{1} = \mathbf{1}$ is said to be a *stochastic matrix*. With these new concepts, we can restate properties (Pa)-(Pb)-(Pc) above by saying that P is a stochastic matrix adapted to G.

As already noticed, P behaves as a local averaging operator: Given a vector $x \in \mathbb{R}^V$, the component v of Px is a weighted average of the values x_w for $w \in \mathcal{N}_v^{\text{out}}$. There is also an interesting *flux* interpretation of the adjoint operator. Given $\zeta \in \mathbb{R}^V$, $(\zeta^* P)_v = \sum \zeta_w P_{wv}$ can be interpreted as follows: From each node w, the quantity ζ_w will flow through the outgoing edges splitting according to the weights P_{wv} as v varies among the out neighbors of w. Hence, $\sum \zeta_w P_{wv}$ is the total new quantity present at node v.

Moreover, a stochastic matrix is the main ingredient of a Markov chain, a special stochastic process such that the future only depends on the past through the present state and states are finite in number. Given a stochastic matrix $P \in \mathbb{R}^{V \times V}$, the term P_{vw} can be interpreted as the probability of making a transition from state v to state w: If you associate each state with the node of the associated graph $\mathcal{G}_P$, you can imagine to be sitting at state v and to walk along one of the available outgoing edges from v according to the various probabilities P_{vw}. In this way, you construct what is called a random walk on the graph G. In this probabilistic setting, flows can be interpreted as probabilities: If $\zeta \in \mathbb{R}^V$ is a probability vector where ζ_v indicates the probability that at the initial instant the state is equal to v, then $(\zeta^* P)_v$ indicates the probability of finding the process in state v at the next time.

The first general observation to be done on stochastic matrices is that the set of stochastic matrices is closed under a number of important operations (whose elementary proof is left to the reader):

(1) If $P, Q \in \mathbb{R}^{V \times V}$ are stochastic, then $\lambda P + (1 - \lambda)Q$ is stochastic for any $\lambda \in (0, 1)$.
(2) If $P, Q \in \mathbb{R}^{V \times V}$ are stochastic, then PQ is stochastic. In particular, P^t is stochastic, for any $t \in \mathbb{N}$.
(3) If P_n is a sequence of stochastic matrices such that $P_n \to P$ for $n \to +\infty$, then P is stochastic.

Properties (1) and (3) say that the set of stochastic matrices form a compact convex subset of $[0, 1]^{V \times V}$.

We are now almost ready to state and prove the main result of this chapter, which investigates the behavior of the powers of a stochastic matrix, proposing minimal conditions to get convergence. The proof is based on the following lemma, which shall also be used later in these notes.

Lemma 2.1 (Contraction principle) *Let $Q \in \mathbb{R}^{V \times V}$ be a stochastic matrix such that there exist $\alpha > 0$ and $m \in V$ such that $Q_{vm} \geq \alpha$ for all $v \in V$. Then, for all $x \in \mathbb{R}^V$, it holds true that $y = Qx$ satisfies*

$$\max_{v \in V} y_v - \min_{v \in V} y_v \leq (1 - \alpha)\big(\max_{v \in V} x_v - \min_{v \in V} x_v\big)$$

Proof Note that

$$\begin{aligned} y_v = \sum_{w\in V} Q_{vw}x_w = &\sum_{w\in V} Q_{vw}(x_w - \min_{u\in V} x_u) + \sum_{w\in V} Q_{vw} \min_{u\in V} x_u \\ \geq &\alpha(x_m - \min_{u\in V} x_u) + \min_{u\in V} x_u \\ = &\alpha x_m + (1-\alpha) \min_{u\in V} x_u. \end{aligned}$$

Similarly,

$$\begin{aligned} y_v = \sum_{w\in V} Q_{vw}x_w = &\sum_{w\in V} Q_{vw}(x_w - \max_{u\in V} x_u) + \sum_{w\in V} Q_{vw} \max_{u\in V} x_u \\ \leq &\alpha(x_m - \max_{u\in V} x_u) + \max_{u\in V} x_u \\ = &\alpha x_m + (1-\alpha) \max_{u\in V} x_u. \end{aligned}$$

Putting these two inequalities together gives:

$$\begin{aligned} \max_{u\in V} y_u - \min_{u\in V} y_u \leq &\alpha x_m + (1-\alpha) \max_{u\in V} x_u - \alpha x_m - (1-\alpha) \min_{u\in V} x_u \\ = &(1-\alpha)(\max_{u\in V} x_u - \min_{u\in V} x_u), \end{aligned}$$

that is the thesis. □

The main result is then the following.

Theorem 2.2 (Convergence to consensus) *Let $P \in \mathbb{R}^{V\times V}$ be a stochastic matrix such that $\mathcal{G}_P$ admits a globally reachable aperiodic vertex. Then, the following two equivalent facts hold true.*

(i) The dynamics (2.4) is such that, for any initial condition $x(0) = x^0 \in \mathbb{R}^V$, there exists a scalar α such that

$$x(t) = P^t x(0) \to \alpha \mathbf{1} \qquad t \to +\infty.$$

In other terms, all components $x_v(t)$ converge to the same consensus value α.

(ii) There exists a vector $\pi \in \mathbb{R}^V$ such that $\pi_v \geq 0$ for all v, $\sum_v \pi_v = 1$, and

$$\lim_{t\to+\infty} P^t = \mathbf{1}\pi^*. \tag{2.7}$$

In other terms, P^t converges to a matrix having all rows equal to the row vector π^*.

Furthermore, $\alpha = \pi^* x(0)$.

Let $s \in V$ be the aperiodic vertex which is reachable from all others. This means that there exists $t^\star \in \mathbb{N}$ such that $Q := P^{t^\star}$ is such that $Q_{vs} > 0$ for all $v \in V$. Let $\alpha = \min\{Q_{vs} : v \in V\} > 0$. Then, letting $y^0 \in \mathbb{R}^V$ and $y^1 = Qy^0$, Lemma 2.1 implies that

$$\max_{v \in V} y^1_v - \min_{v \in V} y^1_v \le (1-\alpha)(\max_{v \in V} y^0_v - \min_{v \in V} y^0_v).$$

Fix now $x(0) = x^0 \in \mathbb{R}^V$ arbitrarily and consider (2.4). Define $M_t = \max_{v \in V} x_v(t)$ and $m_t = \min_{v \in V} x_v(t)$, and notice that, since the components of $x(t)$ are convex combinations of those of $x(t-1)$, the sequences M_t and m_t are bounded and, respectively, nonincreasing and nondecreasing (hence convergent). Hence, also $\Delta_t = M_t - m_t$ converges. For the previous argument, moreover, it holds that $\Delta_{nt^\star} \le (1-\alpha)^n \Delta_0$. This implies that $\Delta_{nt^\star} \to 0$ for $n \to +\infty$. Hence, all components of $x(t)$ will converge to the same limit, thus proving the first claim. If we apply this result choosing $x(0) = e_w$, the wth element of the canonical basis of $\mathbb{R}^V$, we obtain that all elements of the wth column of P^t will converge to the same limit. This clearly yields the second claim.

□

Theorem 2.2 immediately yields Theorem 2.1. An important special case is discussed in the following remark.

Remark 2.1 (*Irreducibility*) A stochastic matrix P for which $\mathcal{G}_P$ is strongly connected is called *irreducible*. A stochastic matrix is said to be aperiodic if $\mathcal{G}_P$ is aperiodic. Hence, Theorem 2.2 applies to the important case when P is irreducible and aperiodic. Notice that for symmetric P these two properties are equivalent to the assumptions in Theorem 2.2.

We now briefly discuss the *necessity* of the assumptions in Theorem 2.2.

Remark 2.2 (*Aperiodicity*) Notice that it is not necessary that all units have access to their own state. It is instead sufficient that the globally reachable node is aperiodic; hence, for instance, it is sufficient that there is a self-loop in this node. The fact that some assumption of aperiodicity is necessary for convergence follows by considering the simple example of a strongly connected graph with two nodes and no self-loops. The only possible stochastic matrix adapted to such a graph is

$$P = \begin{pmatrix} 0 & 1 \\ 1 & 0 \end{pmatrix}$$

Notice that $P^{2t} = I$ for all t, and therefore, P does not yield consensus. A few properties of periodic matrices are discussed in Exercise 2.3.

Remark 2.3 (*Connectivity*) If the graph $\mathcal{G}_P$ has two (or more) sinks, the matrix P can be written with the block structure

$$P = \begin{bmatrix} R_1 & R_2 & R_3 \\ 0 & P_1 & 0 \\ 0 & 0 & P_2 \end{bmatrix}$$

Since the powers of P inherit its block structure, the entries of no column of P^t can converge to the same value in general. This reasoning shows that *global reachability of a node is necessary for convergence to consensus.*

The convergence theorem is illustrated by the following example and in Fig. 2.1.

Example 2.1 (*An irreducible, aperiodic stochastic matrix*) Consider the stochastic matrix

$$P = \begin{bmatrix} 1/2 & 1/2 & 0 \\ 1/3 & 1/3 & 1/3 \\ 1/3 & 0 & 2/3 \end{bmatrix}$$

It is evident that P is irreducible and aperiodic. Let us compute the invariant probability π. From $\pi^* P = \pi^*$, we get

$$\begin{cases} -\frac{1}{2}\pi_1 + \frac{1}{3}\pi_2 + \frac{1}{3}\pi_3 = 0 \\ \frac{1}{2}\pi_1 - \frac{2}{3}\pi_2 \qquad\qquad = 0 \\ \frac{1}{3}\pi_2 - \frac{1}{3}\pi_3 \qquad\qquad = 0 \end{cases}$$

which immediately yields $\pi_2 = \pi_3$ and $\pi_1 = \frac{4}{3}\pi_2$. Using the normalization condition $\pi_1 + \pi_2 + \pi_3 = 1$, we finally get $\pi = \left(\frac{2}{5}, \frac{3}{10}, \frac{3}{10}\right)^*$.

Theorem 2.2 also contains further information useful to address issue (b) presented at the beginning of the chapter. We shall make this information explicit in the next two results, as well as in the following sections. The first result is about the spectrum of the update matrix.

Corollary 2.1 *Let $P \in \mathbb{R}^{V \times V}$ be a stochastic matrix such that $\mathcal{G}_P$ admits a globally reachable aperiodic node. Then,*

(i) 1 is an algebraically simple eigenvalue whose eigenspace is generated by $\mathbf{1}$;

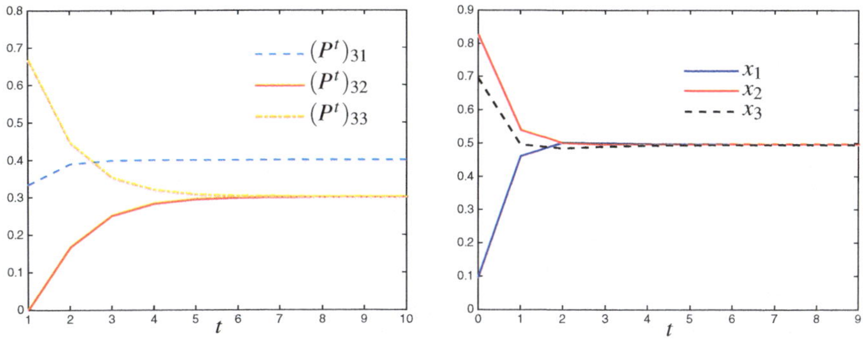

Fig. 2.1 Illustration of convergence for Example 2.1. The *left* diagram plots the entries of the third row of P^t that converges to π^*. The *right* diagram plots the associated averaging dynamics (2.4) from random initial conditions within (0, 1)

(ii) Any other eigenvalue μ of P is such that $|\mu| < 1$.

Proof Suppose indeed that P satisfies the assumptions of Theorem 2.2 and let $\xi \in \mathbb{R}^V$ be an eigenvector of P with eigenvalue μ. Then, for $t \to +\infty$,

$$\mu^t \xi = P^t \xi \to \mathbf{1}\pi^* \xi$$

This immediately yields that either $\mu = 1$ and ξ is a multiple of $\mathbf{1}$, or $|\mu| < 1$. This remark yields (ii) and says that 1 is a geometrically simple eigenvalue (the corresponding eigenspace has dimension 1). It remains to show that 1 is also algebraically simple. This follows using similar arguments showing that the presence of a nontrivial Jordan block relative to the eigenvalue 1 will imply that P^t would grow unbounded contrarily to what is asserted in Theorem 2.2. □

The second result further investigates the structure of the limit matrix.

Corollary 2.2 *Let $P \in \mathbb{R}^{V \times V}$ be a stochastic matrix such that $\mathcal{G}_P$ admits a globally reachable aperiodic node. Consider the vector π as in Theorem 2.2. Then, $\pi^* P = \pi^*$, and π is the only vector sharing this property and the normalization condition $\sum_v \pi_v = 1$.*

Proof A very well-known fact of linear algebra says that P and P^* have the same eigenvalues. This implies that there must exists $\zeta \in \mathbb{R}^V$ such that $\zeta^* P = \zeta^*$. This yields, for $t \to +\infty$,

$$\zeta^* = \zeta^* P^t \to \zeta^* \mathbf{1}\pi^*$$

Hence, ζ is necessarily a multiple of π. In other words, this shows that π is a left eigenvalue of P relative to the eigenvalue 1. Since 1 is also algebraically simple as a left eigenvalue, the uniqueness result immediately follows. □

In the flux interpretation presented at the beginning of this section, the equation $\pi^* P = \pi^*$ can be interpreted as a "stationary regime": The flux is not modifying the quantity π_v present in every node. For this reason, and because of the normalization to 1, π is called *stationary* or *invariant probability measure*. Note that the invariant probability measure needs not to be unique in general—find an example as an exercise—, but is unique when there is a globally reachable node; see Exercise 2.3.

2.4 Convergence Rate and Eigenvalues

This section deals more precisely with question (b1) defined at the beginning of this chapter, that is with convergence speed. The speed of convergence of (2.7) is dictated by the magnitude of the eigenvalues of P. We start by recalling the following result, which is a standard fact in the stability of linear dynamical systems.

Lemma 2.2 *Let $M \in \mathbb{R}^{V \times V}$ be any matrix and let λ_i be its eigenvalues. Let $\rho = \max |\lambda_i|$ be the spectral radius of M. Then, for every $\epsilon > 0$, there exists a constant C_ϵ such that*

$$||M^t x_0||_2 \leq C_\epsilon (\rho + \epsilon)^t ||x_0||_2 \quad \textit{for all } t.$$

A simple application of this lemma allows us to obtain the following result:

Proposition 2.2 (Convergence rate) *Let $P \in \mathbb{R}^{V \times V}$ be a stochastic matrix such that $\mathcal{G}_P$ admits a globally reachable aperiodic node. Consider all its eigenvalues μ_i but 1 and put $\rho_2 = \max\{|\mu_i| < 1\}$. Then, for every $\epsilon > 0$, there exists a constant C_ϵ such that*

$$||(P^t - \mathbf{1}\pi^*)x_0||_2 \leq C_\epsilon (\rho_2 + \epsilon)^t ||x_0||_2 \quad \textit{for all } t.$$

Proof Put $Q := P - \mathbf{1}\pi^*$, and notice that $Q^t = P^t - \mathbf{1}\pi^*$ (check this for exercise). Notice moreover that $Q\mathbf{1} = 0$. Consider now the subspace W of $\mathbb{R}^V$ orthogonal to the vector π, and notice that if $w \in W$, then $Pw = Qw$ and $\pi^* P w = \pi^* w = 0$. In other terms, the subspace W is invariant for P and Q, and on W, the two matrices P and Q coincide. The eigenvalues of P and Q on W are exactly given by the eigenvalues of P different from 1. Wrapping up, we have that Q has eigenvalues μ_i plus the eigenvalue 0; hence, it is asymptotically stable, and the result follows from Lemma 2.2. □

The parameter ρ_2, introduced in the statement of the corollary above, is also called the *second eigenvalue* of P, and the difference $1-\rho_2$ the *spectral gap* of P. The above result essentially says that convergence to rendezvous happens exponentially fast as ρ_2^t. Actually, this is only approximately true because of the arbitrarily small ϵ we have to fix. The ϵ is needed because of the possible presence in P of nontrivial Jordan blocks (which is when the algebraic and geometric dimension of an eigenspace does not coincide). When P is symmetric, things are much simpler: We can indeed follow the proof of Equation (2.6) above and prove the following result which extends Proposition 2.1 to general symmetric matrices P, possibly adapted to nonregular graphs.

Corollary 2.3 (Convergence rate for symmetric P) *Let $P \in \mathbb{R}^{V \times V}$ be a symmetric stochastic matrix such that $\mathcal{G}_P$ is strongly connected and aperiodic. Then,*

$$||(P^t - N^{-1}\mathbf{1}\mathbf{1}^*)x_0||_2 \leq \rho_2^t ||x_0||_2$$

Example 2.2 (*An irreducible, aperiodic stochastic matrix*) Consider the stochastic matrix P defined in Example 2.1. An easy computation shows that the characteristic polynomial of P is given by

$$p(\lambda) := \det(\lambda I - P) = (\lambda - 1)(\lambda - 1/6)^2$$

Therefore, $\rho_2 = 1/6$.

Once we have established that ρ_2 is the right parameter to analyze the speed of convergence, it remains to understand how it depends on the graph: This analysis will be done, on certain families, in Sect. 2.7.

2.5 Consensus Point

Another important question—mentioned in (b2) at the beginning of this chapter—regards the location of the consensus point with respect to the initial condition. As we know from Theorem 2.2, this is completely determined by the left eigenvector π of P. First, notice that the optimization problem:

$$\min_{y \in \mathbb{R}} \sum_{v \in V} |y - y_v|^2$$

has solution given by the barycenter $y = N^{-1} \sum_v y_v$. The rendezvous problem has the barycenter as meeting point if and only if $\pi = N^{-1}\mathbf{1}$. When will this happen? The answer is very simple: if and only if $\mathbf{1}^* P = \mathbf{1}^*$, namely if all columns of P also sum up to 1. When this happens, P is called a *doubly stochastic* matrix. A particular case is when P is symmetric.

What about the possibility to construct a doubly stochastic matrix over a preassigned graph? Is that always possible? The answer is on the negative. Before showing this fact, we introduce another concept which will also be useful later on.

Definition 2.1 (*Sub-stochastic matrix*) A nonnegative matrix $P \in \mathbb{R}^{V \times V}$ is said to be *sub-stochastic* if $\sum_w P_{vw} \leq 1$ for all $v \in V$, and there exists at least one $v \in V$ for which the inequality is strict. Such node will be called a *deficiency node* of P.

There are a few useful facts about sub-stochastic matrices, which the reader is encouraged to verify on his/her own and which are gathered in the following proposition.

Proposition 2.3 (Sub-stochastic matrices) *Let $P \in \mathbb{R}^{V \times V}$ be a sub-stochastic matrix.*

- *Then, P^t is sub-stochastic for all t. More precisely, if we let $V_t^\star$ to be the set of deficiency nodes of P^t, then*

$$V_t^\star \subseteq V_{t+1}^\star \quad \textit{for all } t \in \mathbb{N}.$$

- *If, moreover, P is such that from every node v there is a path in $\mathcal{G}_P$ to a deficiency node, then there exists $t^\star$ such that $V_{t^\star}^\star = V$ (all nodes for $P^{t^\star}$ are deficiency nodes).*

Actually, this fact implies a simple condition for the stability of sub-stochastic matrices.

Proposition 2.4 (Stability of sub stochastic matrices) *Let $P \subset \mathbb{R}^{V \times V}$ be a sub-stochastic matrix such that from every node v there is a path in $\mathcal{G}_P$ to a deficiency node. Then, P is asymptotically stable.*

Proof Let $t^\star$ be defined as in Proposition 2.3 and let $\nu = \max_v \sum_w P^{t^\star}_{vw} < 1$. Given any $t \in \mathbb{N}$, write $t = nt^\star + r$ with $r \in \{0, \ldots, t^\star - 1\}$ and $n \in \mathbb{N}$, and notice that $P^t\mathbf{1} \leq P^{nt^\star}\mathbf{1} \leq \nu^n\mathbf{1}$ (where inequalities have to be understood componentwise). This inequality implies that P^t converges to 0. □

The following result characterizes the "zero pattern" of the invariant probability measure of stochastic matrices and shows that it is not always possible to construct a doubly stochastic matrix on a given graph.

Proposition 2.5 (Positivity of invariant probability measure) *Let $P \in \mathbb{R}^{V \times V}$ be a stochastic matrix such that $\mathcal{G}_P$ admits a globally reachable node $v^\star$. Let π be its invariant probability measure. Then, $\pi_v \neq 0$ if and only if v and $v^\star$ are in the same strongly connected component of $\mathcal{G}_P$.*

Proof Let $V^\star$ be the set of nodes corresponding to the connected component containing $v^\star$, and let $V^{\star\star} = V \setminus V^\star$. Ordering nodes in such a way that the first ones are those in $V^{\star\star}$, we get that P has the following block structure

$$P = \begin{bmatrix} Q & R \\ 0 & S \end{bmatrix}$$

where $Q \in \mathbb{R}^{V^{\star\star} \times V^{\star\star}}$, $R \in \mathbb{R}^{V^{\star\star} \times V^\star}$, and $S \in \mathbb{R}^{V^\star \times V^\star}$. By the assumption made, it follows that Q is sub-stochastic satisfying the assumptions of Proposition 2.4. On the other hand, P^t will have the following block structure:

$$P^t = \begin{bmatrix} Q^t & R_t \\ 0 & S^t \end{bmatrix}.$$

If we partition accordingly $\pi = (\pi^{\star\star}, \pi^\star)$, we then obtain $(\pi^{\star\star})^* Q^t = \pi^{\star\star}$ for all t. This yields $\pi^{\star\star} = 0$. We now prove that instead $\pi^\star_v > 0$ for every $v \in V^\star$. Assume, by contradiction, that there exists $w \in V^\star$ such that $\pi_w = 0$. The relation $\sum_{v \in V^\star} \pi_v P_{vw} = 0$ yields $\pi_v = 0$ for every $v \in N^{in}_w$. A straightforward inductive argument now shows that $\pi_v = 0$ for all $v \in V^*$ for which there exists a path from v to w. By the definition of V^*, this implies that $\pi_v = 0$ for all $v \in V^*$. But this says that π is the 0 vector and can not be an invariant probability. The proof is thus complete. □

This result implies that for a matrix to be doubly stochastic, its associated graph must be strongly connected.

2.6 Stochastic Matrices Adapted to a Graph

In this section, we focus on the problem of finding, given a graph, a stochastic/doubly stochastic matrix adapted to it. One solution is based on assigning equal weight to all outgoing edges of a node, similarly to what we did in (2.5):

$$P = D_G^{-1} A_G. \tag{2.8}$$

This matrix is known as the *simple random walk* (SRW) matrix associated with G. This name is explained by a probabilistic interpretation. Let us think of token that is performing a random walk on the nodes of the graph, according to the following rule: From each node, transitions happen with equal probability along all the available edges. Then, the rows of the SRW matrix are the transition probabilities from each node. More generally, we can consider, for $\tau \in (0, 1)$,

$$P = (1 - \tau)I + \tau D_G^{-1} A_G,$$

which is also stochastic and is called the *lazy* SRW. If G contains a globally reachable aperiodic vertex, then $D_G^{-1} A_G$ yields consensus. On the other hand, even if the globally reachable vertex of G is not aperiodic, the lazy SRW yields consensus for any $\tau \in (0, 1)$ since aperiodicity is automatically gained from the presence of the identity part. Notice that P is not symmetric even when G is symmetric (unless G is also regular which was the case studied in Sect. 2.2). However, the case when G is symmetric is very special since in this case it is very simple to compute the invariant probability measure as $\pi_v = d_v/d$ where $d = \sum_u d_u$. Indeed,

$$(\pi^* P)_w = \sum_v \pi_v P_{vw} = \sum_{v \in \mathcal{N}_w} \frac{d_v}{d} \frac{1}{d_v} = \frac{|\mathcal{N}_w|}{d} = \pi_w.$$

Notice that the invariant measure for the SRW (or for the lazy version) is the uniform vector $N^{-1}\mathbf{1}$ if and only if G is regular. For symmetric nonregular graphs, there is however an alternative construction yielding a symmetric stochastic matrix. It is sufficient to define, for any $v \neq w$,

$$P_{vw} := (A_G)_{vw} \min \left\{ \frac{1}{d_v}, \frac{1}{d_w} \right\}.$$

It is easy to see that, with this choice, the off-diagonal terms of any row of P sum up to a value which is not greater than 1. To complete, we define P on the diagonal terms in such a way to make it a stochastic matrix. Notice that P is symmetric by construction and is called the *Metropolis random walk*.Next, we present an example of Metropolis random walk.

Example 2.3 (*Random walks on a symmetric graph*) Let G be the graph represented in Fig. 2.2. This graph is not regular, and its degree matrix is $D = \text{diag}(2, 2, 2, 3, 1)$.

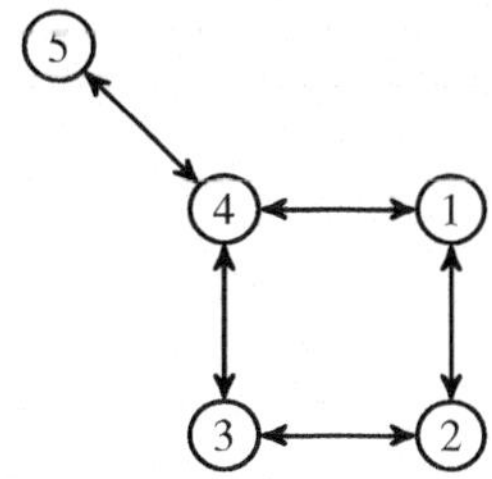

Fig. 2.2 The graph of Example 2.3

Then, the matrix corresponding to a SRW on G is

$$P = \begin{bmatrix} 1/2 & 0 & 0 & 0 & 0 \\ 0 & 1/2 & 0 & 0 & 0 \\ 0 & 0 & 1/2 & 0 & 0 \\ 0 & 0 & 0 & 1/3 & 0 \\ 0 & 0 & 0 & 0 & 1 \end{bmatrix} \begin{bmatrix} 0&1&0&1&0 \\ 1&0&1&0&0 \\ 0&1&0&1&0 \\ 1&0&1&0&1 \\ 0&0&0&1&0 \end{bmatrix} = \begin{bmatrix} 0 & 1/2 & 0 & 1/2 & 0 \\ 1/2 & 0 & 1/2 & 0 & 0 \\ 0 & 1/2 & 0 & 1/2 & 0 \\ 1/3 & 0 & 1/3 & 0 & 1/3 \\ 0 & 0 & 0 & 1 & 0 \end{bmatrix}.$$

Note that this matrix is not doubly stochastic and has -1 as an eigenvalue, because G is bipartite. Instead, the transition matrix of the Metropolis RW is

$$Q = \begin{bmatrix} 1/6 & 1/2 & 0 & 1/3 & 0 \\ 1/2 & 0 & 1/2 & 0 & 0 \\ 0 & 1/2 & 1/6 & 1/3 & 0 \\ 1/3 & 0 & 1/3 & 0 & 1/3 \\ 0 & 0 & 0 & 1/3 & 2/3 \end{bmatrix},$$

which is doubly stochastic and has second eigenvalue $\rho_2(Q) \simeq 0.7845$.

Metropolis construction produces a doubly stochastic matrix, but works for symmetric graphs only. On which graphs is it possible to construct a doubly stochastic matrix adapted to a generic graph? The following result gives us the answer.

Proposition 2.6 (Existence of doubly stochastic P) If $G = (V, E)$ is strongly connected, then there exists a doubly stochastic $P \in \mathbb{R}^{V \times V}$ such that $\mathcal{G}_P = G$.

Proof Given any circuit in G with edges

$$E' = \{(k_1, k_2), (k_2, k_3), \dots, (k_n, k_1)\} \qquad (k_i \neq k_j \text{ for } i \neq j),$$

consider the matrix $P^{(E')} \in \mathbb{R}^{V \times V}$ defined by

$$P^{(E')}_{vw} = \begin{cases} 1 & \text{if } (v, w) \in E' \\ 1 & \text{if } v = w \neq k_s \ \forall s = 1, \dots, n \\ 0 & \text{otherwise} \end{cases}$$

It is immediate to check that P has the following property: On each row and on each column, there is exactly one entry equal to 1, while all the others are equal to 0. This is what is called a *permutation matrix*, a very special case of doubly stochastic matrix. Now consider the family $\mathscr{D}$ of all possible circuits and the convex combination $P = \frac{1}{|\mathscr{D}|}\sum_{E' \in \mathscr{D}} P^{(E')}$: Clearly P is doubly stochastic and $\mathscr{G}_P \subseteq G$. Since G is strongly connected, any edge in G belongs to at least one of the subgraphs in $\mathscr{D}$ (check this as an exercise): This fact implies that $\mathscr{G}_P = G$ and the proof is complete. □

Example 2.4 (*Doubly stochastic matrix*) Consider the graph $G = (V, E)$ with $V = \{1, 2, 3\}$ and $E = \{(1, 2), (2, 3), (3, 1), (2, 1)\}$. Graph G is strongly connected and we know from the proof of Proposition 2.6 that a doubly stochastic matrix can be constructed as

$$P = \frac{1}{2}\left(\begin{bmatrix} 0 & 1 & 0 \\ 0 & 0 & 1 \\ 1 & 0 & 0 \end{bmatrix} + \begin{bmatrix} 0 & 1 & 0 \\ 1 & 0 & 0 \\ 0 & 0 & 1 \end{bmatrix}\right) = \begin{bmatrix} 0 & 1 & 0 \\ 1/2 & 0 & 1/2 \\ 1/2 & 0 & 1/2 \end{bmatrix}.$$

The proof of Proposition 2.6 provides a method to construct a doubly stochastic matrix: The method can be easily applied on small graphs, but is not suitable to large graphs because one needs to find all possible circuits in G. In view of this difficulty, we become interested in matrices whose dominant left eigenvector is not $\mathbf{1}$, but some other vector that still guarantees a "balanced" consensus point. The following definitions go in this direction, identifying sequences of matrices that do not give "too much" weight to any node. Consider a sequence of irreducible matrices $\{Q_n\}_{n\in\mathbb{N}}$ of increasing size, together with their unique invariant measures $\pi^{(n)}$, such that $Q_n^*\pi^{(n)} = \pi^{(n)}$ and $\mathbf{1}^*\pi^{(n)} = 1$. We say that Q_n is *democratic* if $\|\pi^{(n)}\|_\infty \to 0$ as $n \to +\infty$. Clearly, a sequence of doubly stochastic matrices is democratic. A less trivial example is a SRW on a bidimensional grid $L_n \times L_m$. More in general, a family of SRWs on undirected graphs G_N is democratic if and only if $\max_v d_v / \sum_u d_u$ goes to zero as $N = |V|$ goes to infinity. This sufficient condition is not satisfied on star graphs, and indeed, the lazy SRW on S_n is

$$P_{\mathrm{SRW}}(\tau) = \begin{bmatrix} 1-\tau & \tau/n\mathbf{1}_n^* \\ \tau\mathbf{1}_n & (1-\tau)I_n \end{bmatrix},$$

which is not democratic. However, a democratic sequence on star graphs can be constructed as

$$P_{\mathrm{dem}} = \begin{bmatrix} 0 & \frac{1}{n}\mathbf{1}_n^* \\ \frac{1}{n}\mathbf{1}_n & (1-\frac{1}{n})I_n \end{bmatrix}.$$

Other examples of democratic matrices are given in the Exercises.

2.7 Convergence Rate: Examples

We have already seen in Proposition 2.2 that the spectral radius of P determines the rate of convergence to consensus. In this section, the spectral radius is studied in some examples and its connection with convergence time is made more explicit.

We first present an example of a family of SRWs on a simple graph.

Example 2.5 (*SRW on cycles*) Let $G = C_n$ be the symmetric cycle graph with n vertices. Since C_n is 2-regular, we have that the matrix of the SRW is

$$P = \mathrm{circ}([0, 1/2, 0, \dots, 0, 1/2]).$$

The eigenvalues of P can be computed as we did for the Laplacian eigenvalues of C_n in Example 1.5. Namely, the eigenvalues of P are

$$\mu_k(P) = \cos\left(\frac{2\pi}{n}k\right).$$

Note that -1 is an eigenvalue if and only if n is even: This corresponds to the graph being bipartite. If instead n is odd, the second eigenvalue of P is

$$\rho_2 = \max\left\{\left|\cos\left(\frac{2\pi}{n}\right)\right|, \left|\cos\left(\frac{2\pi}{n}\frac{n-1}{2}\right)\right|\right\} = \cos\frac{\pi}{n}.$$

In order to control the convergence properties of a matrix adapted to C_n, we may define the family of matrices $P_\tau = (1-\tau)I + \tau P$, with the parameter $\tau \in (0, 1]$. A matrix in this family corresponds to a lazy SRW. In this case, $\mu_k(\tau; P) = 1 - \tau + \tau\cos\left(\frac{2\pi}{n}k\right)$ and

$$\rho_2(\tau) = \begin{cases} \max\{|1-\tau+\tau\cos\frac{2\pi}{n}|, |1-2\tau|\} & \text{if } n \text{ is even} \\ \max\{|1-\tau+\tau\cos\frac{2\pi}{n}|, |1-\tau(1+\cos\frac{\pi}{n})|\} & \text{if } n \text{ is odd.} \end{cases}$$

It is clear that a choice of $\tau \in (0, 1)$ allows to have ρ_2 smaller than 1 for every n. Furthermore, if n is even,

$$\rho_2(\tau) = \begin{cases} 1-\tau(1-\cos\frac{2\pi}{n}) & \text{if } \tau \le \frac{2}{3-\cos\frac{2\pi}{n}} \\ 2\tau - 1 & \text{otherwise} \end{cases}$$

and the minimum is achieved for $\tau = \frac{2}{3-\cos\frac{2\pi}{n}}$. If instead n is odd,

$$\rho_2(\tau) = \begin{cases} 1-\tau(1-\cos\frac{2\pi}{n}) & \text{if } \tau \le \frac{2}{2+\cos\frac{\pi}{n}-\cos\frac{2\pi}{n}} \\ \tau(1+\cos\frac{\pi}{n}) - 1 & \text{otherwise.} \end{cases}$$

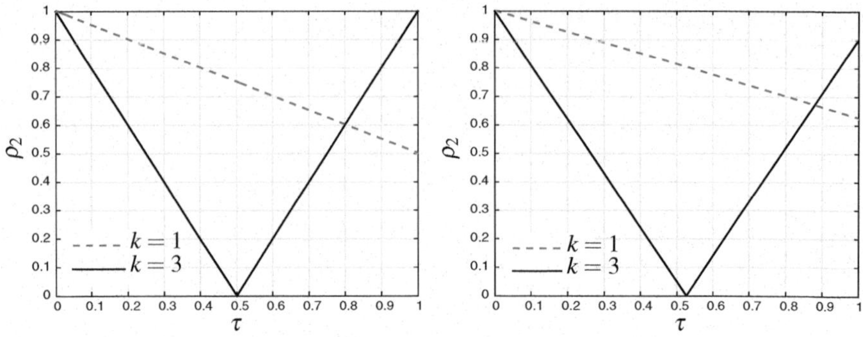

Fig. 2.3 Optimization in Example 2.5. Plots show the function $|\mu_k(\tau; P)|$ for $k=1$ and $k=\lfloor \frac{n}{2} \rfloor$ assuming $n=6$ (*left plot*) and $n=7$ (*right plot*), respectively

and the minimum is achieved for $\tau = \frac{2}{2+\cos\frac{\pi}{n}-\cos\frac{2\pi}{n}}$. This optimization is illustrated in Fig. 2.3

The simple random walk can be also easily studied on the complete graph.

Example 2.6 (*SWR on complete graphs*) The SWR matrix on the complete graph is $P = \frac{1}{N-1}(\mathbf{1}\mathbf{1}^* - I)$, resulting in $\rho_2 = \frac{1}{N-1}$. If we instead consider the lazy version $P(\tau) = (1-\tau)I + \tau P$, we obtain $\rho_2(\tau) = 1 - \frac{N}{N-1}\tau$, which vanishes when $\tau = 1 - \frac{1}{N}$. Indeed, $P(1-\frac{1}{N}) = \frac{1}{N}\mathbf{1}\mathbf{1}^*$ gives consensus in one step.

As we have observed in Example 2.5, -1 is an eigenvalue of the SRW matrix when the graph is bipartite. To rule out such undesired case, in the next example we concentrate on a specific class of lazy SRW.

Example 2.7 (*Spectral radius of k-dimensional grids*) Let G be a d-regular graph, and consider the matrix $P = \frac{1}{d+1}(I+A)$. When the spectrum of A is known, ρ_2 can be readily computed. For instance, for k-dimensional symmetric grids C^k, it holds

$$\rho_2 = \frac{2k-1}{2k+1} + \frac{2}{2k+1}\cos\frac{2\pi}{n}.$$

If the number of nodes $N = n^k$ goes to infinity while keeping the dimension k fixed, then $\rho_2 \to 1$, and by the Taylor expansion of the cosine, we observe that[1]

$$\rho_2 = 1 - \frac{4\pi^2}{2k+1}\frac{1}{n^2} + o\left(\frac{1}{n^3}\right) = 1 - \frac{4\pi^2}{2k+1}\frac{1}{N^{2/k}} + o\left(\frac{1}{N^{3/k}}\right) \qquad \text{as } n \to \infty.$$

[1]Here and throughout the book, we will make use of the standard asymptotic notation. If f_n and g_n are two positive sequences, we say that $f_n = O(g_n)$ if f_n/g_n is upper bounded in n; that $f_n = \Theta(g_n)$ if f_n/g_n is both lower and upper bounded in n; that $f_n = o(g_n)$ if $f_n/g_n \to 0$ as $n \to \infty$; and that $f_n \sim g_n$ if $f_n/g_n \to 1$ as $n \to \infty$.

Hence, $1-\rho_2$ goes to zero at a polynomial rate. More details and more graph examples on this SRW are given in Exercise 2.12.

To better highlight the role of N, we define the *convergence time* as

$$T_\varepsilon = \inf\{t > 0 \, : \, ||P^t - \mathbf{1}\pi^*|| < \varepsilon\}.$$

On symmetric matrices, $||P^t - \mathbf{1}\pi^*|| = \rho_2^t$, so that $T_\varepsilon = \frac{\log \varepsilon^{-1}}{\log \rho_2^{-1}}$, which is in turn upper bounded by $\frac{\log \varepsilon^{-1}}{1-\rho_2}$. Moreover, $\frac{1}{1-\rho_2} \sim \frac{1}{\log \rho_2^{-1}}$ as $\rho_2 \to 1$. Hence, the inverse of the spectral gap of P can be immediately interpreted as an upper bound on the convergence time. For instance, in the example above, $T_\varepsilon \sim \frac{2k+1}{4\pi^2} N^{2/k} \log \varepsilon^{-1}$.

Remark 2.4 (*Trade-off between speed and democracy*) Consider again irreducible matrices adapted to star graphs as at the end of Sect. 2.6. You can easily verify—exercise—that P_{dem} has second eigenvalue $\rho_2(P_{\text{dem}}) = 1 - \frac{1}{n}$ (thus growing to 1 as $n \to= \infty$), whereas $\rho_2(P_{\text{SRW}}(\tau)) = 1 - \tau$. On the other hand, P_{dem} is democratic while $P_{\text{SWR}}(\tau)$ is not. This observation highlights that by choosing either matrix we are trading off speed for democracy. This trade-off exhibited by star graphs is further discussed in Exercise 2.18; see also Exercises 2.23 and 2.24 for other graphs having this feature.

2.8 Reversible Matrices

An important family of stochastic matrices, encompassing SRW and in general all symmetric stochastic matrices, is the family of *reversible* matrices. A reversible matrix can be defined starting from any nonnegative symmetric matrix $M \in \mathbb{R}^{V \times V}$ putting

$$P_{vw} = \frac{M_{vw}}{(M\mathbf{1})_v}. \tag{2.9}$$

It is immediate to check that P is stochastic and that $\pi_v = (M\mathbf{1})_v[\sum_u (M\mathbf{1})_u]^{-1}$ is an invariant probability measure of P. Notice that SRW on symmetric graphs is a special case of this construction, when M is the adjacency matrix of the graph. We have the following alternative characterization:

Proposition 2.7 (Reversibility) *Let P be a stochastic matrix. The following conditions are equivalent:*

(i) P is reversible;
(ii) There exists a nonzero and nonnegative $x \in \mathbb{R}^V$ such that

$$x_v P_{vw} = x_w P_{wv} \quad \textit{for all } v, w \in V. \tag{2.10}$$

Proof On the one hand, if P satisfies (2.9), it follows that

$$(M\mathbf{1})_v P_{vw} = M_{vw} = M_{wv} = (M\mathbf{1})_w P_{wv}$$

On the other hand, if P satisfies (2.10), then putting $M_{vw} := x_v P_{vw}$ we have that M is nonnegative symmetric, and it holds $(M\mathbf{1})_v = x_v$. Hence, (2.9) holds with such M. □

Condition (2.10) is often referred to as the *detailed balance* condition: We note that it implies that x is actually an invariant measure (possibly not normalized to 1) of P, because

$$\sum_v x_v P_{vw} = \sum_v x_w P_{wv} = x_v.$$

Condition (2.10) is actually stronger than the requirement that x is an invariant measure as it says that each pair of nodes v, w for which $P_{vw} > 0$ must balance the exchange flow between each other. The reason for the name "reversible" becomes clear when we interpret it in the probabilistic framework considering P as the transition matrix of a Markov chain X_t having initial probability vector π satisfying the condition (2.10). Then, the left and right terms of (2.10) can be interpreted, respectively, as $\mathbb{P}(X_t = v, X_{t+1} = w)$ and $\mathbb{P}(X_t = w, X_{t+1} = v)$.

It is possible to generalize to reversible matrices most of the results obtained for symmetric matrices: The key fact is that reversible matrices are *diagonalizable* as we show below. Let $P \in \mathbb{R}^{V \times V}$ be a reversible, irreducible, aperiodic stochastic matrix, and let $\pi \in \mathbb{R}^V$ be its invariant probability measure. Consider D_π the diagonal matrix such that $(D_\pi)_{vv} = \pi_v$, and define $A = D_\pi^{1/2} P D_\pi^{-1/2}$. Reversibility implies (check this) that A is symmetric. Let ϕ_j's, for $j \in \{1, \ldots, n\}$, be an orthonormal basis of eigenvectors for A with correspondent real eigenvalues μ_j. It is immediate to check that $\pi^{1/2}$ is indeed an eigenvector with eigenvalue 1. Therefore, we will assume that $\phi_1 = \pi^{1/2}$ and $\mu_1 = 1$. A straightforward verification shows that the $\psi_j = D_\pi^{-1/2} \phi_j$ are eigenvectors of P with eigenvalue μ_j. The ψ_j's together with $\mathbf{1}$ do form a basis of eigenvectors of P which is thus diagonalizable. Using the usual orthonormal splitting expression for A, we can write

$$A^t = \pi^{1/2}(\pi^{1/2})^* + \sum_{j \geq 2} \mu_j^t \phi_j \phi_j^*,$$

from which we can derive the following useful representation for P^t

$$P^t = \mathbf{1}\pi^* + D_\pi^{-1/2} \sum_{j \geq 2} \mu_j^t \phi_j \phi_j^* D_\pi^{1/2}.$$

From this expression, we can estimate the speed of convergence as in the symmetric case (see Problem 2.19 for details). Moreover, we can extend the theory developed for the Laplacian $L(P) = I - P$, when P is a symmetric matrix, to the case when

P is reversible. The idea for the extension simply involves replacing the Euclidean scalar product with the product induced by π, which is $\langle x, y\rangle_\pi := < x, D_\pi y > = \sum_v \pi_v x_v y_v$. In particular, the following results, extending Propositions 1.9 and 1.9, hold true (their proof is left to the reader).

Proposition 2.8 (Dirichlet form for reversible matrices) *Assume that P is a reversible stochastic matrix with invariant probability measure π. For every $x \in \mathbb{R}^V$, it holds*

$$\langle x, L(P)x\rangle_\pi = \frac{1}{2}\sum_{v,w} P_{vw}\pi_v(x_v - x_w)^2. \tag{2.11}$$

Proposition 2.9 (Variational characterization for reversible matrices) *Assume that P is a reversible stochastic matrix with invariant probability measure π and second largest eigenvalue μ_2. Let λ_2 be the second smallest eigenvalue of $L(P)$. It holds*

$$\lambda_2 = (1 - \mu_2) = \min_{x\neq 0, \langle x, \mathbf{1}\rangle_\pi = 0} \frac{\langle x, L(P)x\rangle_\pi}{\langle x, x\rangle_\pi}. \tag{2.12}$$

A useful technique to upper bound the spectral gap of a reversible stochastic matrix P is through the so-called *bottleneck ratio*, a sort of index measuring how well the "flow" represented by the matrix is spreading along the underlying graph. Suppose π is the usual invariant probability measure of P, and for every $S \subset V$, define $\pi(S) = \sum_{v\in S}\pi_v$ and

$$Q(S, S^c) = \sum_{v\in S, w\notin S} \pi_v P_{vw}$$

(check as an exercise that $Q(S, S^c) = Q(S^c, S)$ for all $S \subset V$). Then, we define

$$\Phi(S) := \frac{Q(S, S^c)}{\pi(S)}$$

and the bottleneck ratio of P as

$$\Phi_* := \min_{S:\pi(S)\leq \frac{1}{2}} \Phi(S).$$

In the flow interpretation $Q(S, S^c)$ represents the total flow exiting S (assuming we are at the stationary regime), while $\Phi(S)$ the fraction of flow exiting S with respect to the total flow exiting from the nodes in S. We have the following result:

Proposition 2.10 (Cheeger bound) *Let μ_2 be the second largest eigenvalue of a reversible matrix P, and let Φ_* be the bottleneck ratio of P. Then,*

$$1 - \mu_2 \leq 2\Phi_*.$$

Proof Given $S \subseteq V$, consider the vector $\phi \in \mathbb{R}^V$ defined by $\phi_v = \pi(S^c)$ if $v \in S$, and $\phi_v = -\pi(S)$ if $v \in S^c$. Then, from Proposition 2.8 and the detailed balance condition (2.10), it follows that

$$\begin{aligned}
\langle \phi, L(P)\phi \rangle_\pi &= \frac{1}{2}\sum_{v,w} \pi_v P_{vw}(\phi_v - \phi_w)^2 \\
&= \sum_{v \in S, w \notin S} \pi_v P_{vw}(\phi_v - \phi_w)^2 \\
&= \sum_{v \in S, w \notin S} \pi_v P_{vw}(\pi(S) + \pi(S^c))^2 \\
&= \sum_{v \in S, w \notin S} \pi_v P_{vw} = Q(S, S^c)\,.
\end{aligned}$$

On the other hand,

$$\langle \phi, \phi \rangle_\pi = \sum_v \pi_v \phi_v^2 = \sum_{v \in S} \pi_v \pi(S^c)^2 + \sum_{w \notin S} \pi_w \pi(S)^2 = \pi(S)\pi(S^c).$$

From the variational characterization of Proposition 2.9, and assuming $\pi(S) \leq 1/2$, we thus conclude

$$\lambda_2 \leq \frac{\langle \phi, L(P)\phi \rangle_\pi}{\langle \phi, \phi \rangle_\pi} = \frac{Q(S, S^c)}{\pi(S)\pi(S^c)} \leq 2\Phi(S).$$

Since this inequality holds for all S such that $\pi(S) \leq \frac{1}{2}$, the upper bound is proved. □

Notice that, since $\rho_2 \geq \mu_2$, we can also bound the spectral gap by

$$1 - \rho_2 \leq 2\Phi_*.$$

In the case when P is the SRW on a symmetric graph $G = (V, E)$, the bottleneck ratio takes a peculiar form which is convenient to work out:

$$\Phi(S) = \frac{\sum\limits_{v \in S, w \in S^c} \frac{d_v}{|E|}(A_G)_{vw}\frac{1}{d_v}}{\sum\limits_{v \in S} \frac{d_v}{|E|}} = \frac{\sum\limits_{v \in S, w \in S^c} (A_G)_{vw}}{\sum\limits_{v \in S} d_v} \tag{2.13}$$

This equation says that $\Phi(S)$ equals the *fraction of edges which start inside S and end outside S*.

Example 2.8 (*Graphs with a bottleneck*) Given two graphs $G_1 = (V_1, E_1)$, $G_2 = (V_2, E_2)$ and a symmetric set of edges $E_3 \subseteq (V_1 \times V_2) \cup (V_2 \times V_1)$, we can consider the interconnected graph $G = (V_1 \cup V_2, E_1 \cup E_2 \cup E_3)$. Following (2.13), for the SRW on the graph G, we have that the bottleneck can be estimated as

$$\Phi_* \leq \Phi(V_1) = \frac{|E_3|}{2|E_1| + |E_3|}$$

For instance, consider the case when $|V_1| = |V_2| = n$, G_1 and G_2 are both complete, and $|E_3| = 2$ (namely, there is just one edge and its inverse) connecting the two complete graphs (this is called *barbell graph*). Then, $\Phi_* \leq (n(n-1)+1)^{-1}$.

Similar reasonings can be applied to other families of matrices; see for instance Exercise 2.20 on Metropolis random walks. For completeness, we report that also a lower bound on the spectral gap involving the bottleneck ratio can be obtained [30, Theorem 13.14].

Proposition 2.11 (Reverse Cheeger bound) *Let μ_2 be the second largest eigenvalue of a reversible matrix P, and let Φ_* be the bottleneck ratio of P. Then,*

$$1 - \mu_2 \geq \frac{{\Phi_*}^2}{2}.$$

Exercises

Exercises are divided into five groups, respectively, devoted to some basic facts, to the rate of convergence, to the consensus value, to reversible matrices, and to miscellaneous arguments.

First Examples and Concepts

Exercise 2.1 (*SRW example*) Consider the graph G in Fig. 2.4, and let P be the transition matrix relative to the simple random walk on G.

(i) Write P.
(ii) Compute P^9_{13}.
(iii) What is the multiplicity of the eigenvalue 1 of P? Why?
(iv) Is -1 an eigenvalue of P? Is P aperiodic? Why?

Exercise 2.2 (*Consensus example*) Consider the simple random walk (2.8) on the graph $G = (V, E)$ defined in Exercise 1.9, and let $x(t)$ be the evolution of the consensus algorithm associated with the corresponding stochastic matrix.

(i) Prove that $x(t)$ converges to consensus at the value α.
(ii) Find the value of α as a function of the initial condition $x(0)$.
(iii) Find $\bar{t}$ such that $\|x(\bar{t}) - |V|^{-1}\mathbf{1}\| \leq 10^{-2}\|x(0)\|$.

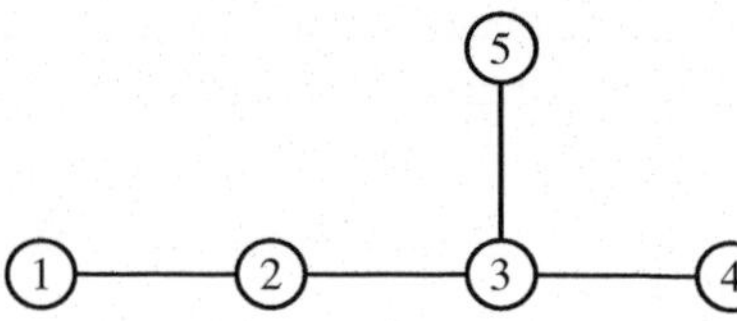

Fig. 2.4 The graph G of Exercise 2.1

Exercise 2.3 (*Periodic matrices*) Let P be a stochastic matrix such that $\mathcal{G}_P$ has a globally reachable node. Prove the following facts.

(i) $Q_\varepsilon = \varepsilon I + (1 - \varepsilon)P$ is stochastic, and $\mathcal{G}_{Q_\varepsilon}$ has a globally reachable aperiodic node.
(ii) The scalar 1 is a simple eigenvalue of P.
(iii) The spectrum of P is contained in the closed unit disk of the complex plane.
(iv) P has a unique invariant probability measure.

Rate of Convergence

Exercise 2.4 (*SRW on complete*) Let G be the complete graph.

(i) Write down explicitly the corresponding symmetric random walk P, as in (2.8).
(ii) Compute all eigenvalues of P and, in particular, the second eigenvalue ρ_2.
(iii) Consider the lazy SRW $P_\tau = (1-\tau)I + \tau P$, and compute the corresponding ρ_2.

Exercise 2.5 (*SRW on complete bipartite*) Consider the complete bipartite graph $K_{m,n}$ as defined in Example 1.2

(i) Write down explicitly the corresponding symmetric random walk P, as in (2.8).
(ii) Compute all eigenvalues of P and, in particular, check that -1 is always an eigenvalue.
(iii) Consider the lazy SRW $P_\tau = (1-\tau)I + \tau P$, and compute the corresponding ρ_2.

Exercise 2.6 (*SRW on grids*) Let $G = C_n \times C_m$ be the symmetric two-dimensional toroidal graph with $n \times m$ vertices.

(i) Observe that the corresponding symmetric random walk P is a Cayley matrix.
(ii) Compute all eigenvalues of P and find when -1 is an eigenvalue.
(iii) When n and m are both odd, compute the second eigenvalue ρ_2 of P.
(iv) Consider the lazy SRW $P_\tau = (1 - \tau)I + \tau P$, and compute the corresponding ρ_2 for every value of n and m.
(v) As τ varies in $[0, 1]$, compute the maximal value of the spectral gap $1 - \rho_2$ (you may assume that n and m are sufficiently large).

Exercise 2.7 (*Symmetric cycle*) Let $n \in \mathbb{N}$, and consider the symmetric cycle graph C_n with adjacency matrix A_n. Consider the matrix $P_n = \frac{1}{3}(I + A_n)$ corresponding to a lazy random walk on C_n.

(i) Verify that P_n is a lazy simple random walk on C_n.

(ii) Let $\rho_2^{(n)}$ be the second largest eigenvalue of P_n. Using the formula for the eigenvalues of circulant matrices in Proposition 1.13, find an expression for $\rho_2^{(n)}$.
(iii) Find a function $f(n)$ such that $f(n) \sim 1 - \rho_2^{(n)}$.

Exercise 2.8 (*Augmented cycle I*) Let $n \in \mathbb{N}$, and consider the following augmentation G_n of the symmetric cycle graph C_n, defined as follows: A node $i \in \{0, \ldots, n-1\}$ is connected with nodes $i-2, i-1, i+1, i+2 \pmod n$. Consider the matrix P_n corresponding to the simple random walk on G_n. Let $\rho_2^{(n)}$ be the second largest eigenvalue of P_n.

(i) Using the formula for the eigenvalues of circulant matrices in Proposition 1.13, find an expression for $\rho_2^{(n)}$.
(ii) Find a function $f(n)$ such that $f(n) \sim 1 - \rho_2^{(n)}$.

Exercise 2.9 (*Augmented cycle II*) Let n be an even number, and consider the following augmentation G_n of the symmetric cycle graph C_n, defined as follows: A node $i \in \{0, \ldots, n-1\}$ is connected with nodes $i-1, i+1, i+n/2 \pmod n$. Consider the matrix P_n corresponding to the simple random walk on G_n. Let $\rho_2^{(n)}$ be the second largest eigenvalue of P_n.

(i) Using the formula for the eigenvalues of circulant matrices, find an expression for $\rho_2^{(n)}$.
(ii) Find a function $f(n)$ such that $f(n) \sim 1 - \rho_2^{(n)}$.

Exercise 2.10 (*Line graph*) Let $n \in \mathbb{N}$, and consider the following matrix P_n corresponding to a random walk on the symmetric line graph L_n:

$$P_n = \frac{1}{2}\begin{bmatrix} 1\,1\ 0 & \ldots & 0\,0 \\ 1\,0\ 1\ 0 & \ldots & 0\,0 \\ 0\,1\ 0\ 1 & \ldots & 0\,0 \\ & \ddots & \\ 0\,0\ldots & 0 & 1\,0 \\ 0\,0\ldots & 1 & 0\,1 \\ 0\,0\ldots & 0 & 1\,1 \end{bmatrix}.$$

Let $\rho_2^{(n)}$ be the second largest eigenvalue of P_n.

(i) Using the formulas for the eigenvalues of tridiagonal matrices in Exercise 1.27, verify that $\rho_2^{(n)} = \cos\frac{\pi}{n}$.
(ii) Consider the simple random walk on a symmetric cycle C_n (see Exercise 2.7), and denote by $\bar{\rho}_2^{(n)}$ the second largest eigenvalue of the associated matrix. Show that $\rho_2^{(n)} \geq \bar{\rho}_2^{(n)}$, and compute $\lim_n \dfrac{1-\rho_2^{(n)}}{1-\bar{\rho}_2^{(n)}}$.
(iii) Interpret the above results in terms of speed of convergence of the corresponding consensus algorithms on L_n and C_n.

Exercise 2.11 (*Line graph II*) Let $n \in \mathbb{N}$, and consider the symmetric line graph L_n with adjacency matrix B_n. Consider the matrix

$$Q_n = \frac{1}{3}(I + B_n) + \frac{1}{3}\begin{bmatrix} 1\,0 \dots 0\,0 \\ 0\,0 \dots 0\,0 \\ \ddots \\ 0\,0 \dots 0\,0 \\ 0\,0 \dots 0\,1 \end{bmatrix}$$

corresponding to a lazy random walk on L_n. Let $\rho_2^{(n)}$ be the second largest eigenvalue of Q_n.

(i) Using Exercise 1.27, find a closed form expression for $\rho_2^{(n)}$.
(ii) Find a function $f(n)$ such that $f(n) \sim 1 - \rho_2^{(n)}$.
(iii) Compare these results with the analogous results for the simple random walk on the cycle graph C_n in Exercise 2.7.

Exercise 2.12 (*Rate comparison*) Consider cycle graphs, k-dimensional torus graphs, k-dimensional hypercubes, and De Bruijn graphs on k symbols.

(i) Observe that the graphs at hand are regular. For each of these graphs, consider its Laplacian matrix L and the stochastic matrix $P = I - \frac{1}{d+1}L = \frac{1}{d+1}(I + A)$. Let ρ_2 be the magnitude of the second largest eigenvalue of P.
(ii) Observe that the graphs at hand have real Laplacian eigenvalues. Denote them as $0 = \lambda_1 < \lambda_2 \leq \cdots \leq \lambda_n$, and verify that $\rho_2 = 1 - \frac{\lambda_2}{d+1}$, where d is the degree of the graph. Observe that under the current assumptions, the rate of convergence of P is completely determined by the topology of the graph.
(iii) Compute the values of ρ_2 as functions of the graph parameters and of the number of nodes N. Derive the values in Table 2.1.
(iv) By using Taylor expansions, compute the principal part of the rate ρ_2 as $N \to +\infty$, in the cases of Table 2.1.
(v) Rank the graphs in Table 2.1 from fastest to slowest. Observe that if we consider a sequence B_h^k with k fixed and $h \in \mathbb{N}$, then ρ_2 does not depend on N.

Exercise 2.13 (*Rate on directed grids*) Let $G = \mathbf{C}_n^d$ be a directed d-dimensional torus, $P = \frac{1}{d+1}(I + A_G)$, and $N = n^d$.

(i) Verify that

$$\rho_2 = \sqrt{\frac{d^2 + 1 + 2d\cos\left(\frac{2\pi}{n}\right)}{(d+1)^2}} = 1 - \frac{2d\pi^2}{(d+1)^2}\frac{1}{n^2} + o\left(\frac{1}{n^3}\right) \text{ as } n \to \infty$$

Exercise 2.14 (*Rate comparison in SRW*) Consider the same graphs as in Exercise 2.12 and for each of them the stochastic matrix $P = \frac{1}{d}A$, corresponding to a symmetric random walk.

Table 2.1 Rates of convergence for consensus algorithms on several families of graphs; see Example 2.7 and Exercise 2.12

Graph	λ_2	d	N	ρ_2	$\rho_2(N)$
C_n	$2(1-\cos\frac{2\pi}{n})$	2	n	$\frac{1}{3}(1+2\cos\frac{2\pi}{n})$	$\frac{1}{3}(1+2\cos\frac{2\pi}{N})$
$C_n \times C_n$	$2(1-\cos\frac{2\pi}{n})$	4	n^2	$\frac{1}{5}(3+2\cos\frac{2\pi}{n})$	$\frac{1}{5}(3+2\cos\frac{2\pi}{\sqrt{N}})$
C_n^k	$2(1-\cos\frac{2\pi}{n})$	$2k$	n^k	$\frac{2k-1}{2k+1}+\frac{2}{2k+1}\cos\frac{2\pi}{n}$	$\frac{2k-1}{2k+1}+\frac{2}{2k+1}\cos\frac{2\pi}{N^{1/k}}$
H_k	2	k	2^k	$\frac{d-1}{d+1}$	$\frac{\log_2 N-1}{\log_2 N+1}$
B_h^k	k	k	k^h	$\frac{1}{k+1}$	$\frac{1}{N^{1/h}+1}$

(i) Compute the second eigenvalues ρ_2 as functions of the graph parameters and of the number of nodes N.
(ii) Compare your results with those in Exercise 2.12.

Exercise 2.15 (*Majority computation*) Let $G = (V, E)$ be a symmetric connected graph and $\bar{x} \in \{-1, +1\}^V$. Let $N_1 = |\{v \in V : \bar{x}_v = 1\}|$ and $N_{-1} = N - N_1$ where $N = |V|$. The agents want to estimate which state has the majority. Consider the following algorithm. Let P be an aperiodic irreducible symmetric matrix adapted to G, and define

$$\begin{cases} x(t) = P^t \bar{x} \\ \lambda(t) = \text{sign}(x(t)) \in \{-1, +1\}^V. \end{cases}$$

Clearly, if $N_1 \neq N_{-1}$, then $\lim_{t\to\infty} \lambda(t) = \bar{\lambda}\mathbf{1}$ and $\bar{\lambda} = 1$ when $N_1 > N_{-1}$. Agent v can then use $\lambda_v(t)$ as an estimation of the majority value. Let $T_{\min} = \min\{t \in \mathbb{N} : \lambda_v(t) = \bar{\lambda}\ \forall v \in V\}$.

(i) Estimate $T_{\min}$ in terms of the second eigenvalue ρ_2 of P and of the vector $\bar{x}$.
(ii) Estimate $T_{\min}$ when P is the SRW in the toroidal d-grid of size N.

Evaluation of the Convergence Value and Democracy

Exercise 2.16 (*Democracy and wisdom of crowds*) Consider an irreducible aperiodic matrix Q used to solve a consensus problem with $x(0) = \theta\mathbf{1} + \eta$, where θ is a scalar and η is a vector of "disturbances." We know that $x_v(t) \to x_v(\infty) = \theta + \pi^*\eta$ for all v. Assume that η_vs are random variables, independent and identically distributed with zero mean and variance σ^2 (this setup will be considered again in Sect. 4.5).

Let $\{Q_N\}_{N\in\mathbb{N}}$ be a sequence of such matrices, each with size N. According to [21], the sequence Q_N is said to be *wise* if the variance of $x_v(\infty)$ goes to 0 as N goes to infinity. Prove the following statements.

(i) Q_N is democratic if and only if it is wise.
(ii) There exists $c > 0$ such that $\frac{\pi_u}{\pi_v} \leq c$ for all indices u, v and all size N if and only if there exist two positive constants c_1 and c_2 such that $\frac{c_1}{N} \leq \pi_w \leq \frac{c_2}{N}$ for all w and all N.
(iii) The conditions at point (ii) imply that Q_N is wise.

Exercise 2.17 (*Line graph: democracy*) Let $n \in \mathbb{N}$, and consider the symmetric line graph L_n. Consider the matrix S_n associated with the simple random walk on L_n, and define

$$Q_n = \frac{1}{3}I + \frac{2}{3}S_n.$$

(i) Observe that Q_n is stochastic but not doubly stochastic, and compute $\pi^{(n)}$, the invariant probability measure of Q_n.
(ii) Compute for each component $v \in \{1, \dots, n\}$,

$$\lim_{n\to+\infty} \pi_v^{(n)} \quad \text{and} \quad \lim_{n\to+\infty} \frac{\pi_v^{(n)}}{1/n}.$$

Comment on your results, recalling that the invariant probability measure of a doubly stochastic matrix is $\frac{1}{n}\mathbf{1}$. Is Q_n democratic?
(iii) Compute $\rho_2(n)$, the second largest eigenvalue of Q_n.
(iv) Compare these figures with the corresponding results for the simple random walk on the cycle graph C_n (see Exercise 2.7).

Exercise 2.18 (*Speed and democracy on star graphs*) Consider the graph S_n, the symmetric star graph with n edges and $n+1$ nodes, whose center node is denoted as 0 and the n leaves as the elements of the set $\{1, \dots, n\}$. Then, consider the following family of adapted stochastic matrices

$$P_n = \begin{bmatrix} 1-n\alpha & \alpha & \alpha & \alpha & \dots & \alpha \\ \beta & 1-\beta & 0 & 0 & \dots & 0 \\ \beta & 0 & 1-\beta & 0 & \dots & 0 \\ & \vdots & & \ddots & & \\ & \vdots & & & & \\ \beta & 0 & \dots & 0 & 0 & 1-\beta \end{bmatrix},$$

where the parameters α, β satisfy $0 \leq \alpha \leq \frac{1}{n}$ and $0 \leq \beta \leq 1$.

(i) Verify that the eigenvalues of P_n are 1, $1-\beta$, and $1-n\alpha-\beta$.
(ii) Find the values of α, β which give the fastest consensus algorithm.

(iii) Design a consensus algorithm, adapted to S_n, which converges in a finite number of steps. How many steps does it need to converge? Compare the required number of steps of the algorithm with the diameter of S_n. Which is the consensus value?
(iv) Compute $\rho_2(n)$ as a function of α, β on its domain.
(v) Compute the invariant probability measure of P_n. Verify that the consensus algorithm defined by P_n converges to the average of the initial states if and only if $\alpha = \beta$.
(vi) Prove that if $\alpha = \beta$, the rate of convergence $\rho_2(n)$ grows to 1 as $n \to +\infty$. Estimate the convergence time on large networks, as a function of n.
(vii) Prove that

(a) if P_n is democratic, then necessarily $\rho_2(n) \to 1$ as n diverges;
(b) conversely, if $\rho_2(n) \leq 1 - c$ for all n and some positive c, then necessarily P_n is not democratic.

Conclude that in optimizing P_n one necessarily needs to trade off the speed of convergence for the distance between the limit value and the average of initial states.

Reversible Matrices

Exercise 2.19 (*Convergence rate*) Suppose $P \in \mathbb{R}^{V \times V}$ is stochastic reversible with invariant probability measure π. Then, the result in Proposition 2.2 can be strengthened to claim that

$$||P^t x(0) - \mathbf{1}\pi^* x(0)||_2 \leq \frac{\max_v \pi_v^{1/2}}{\min_v \pi_v^{1/2}} ||x(0)||_2 \rho_2^t \quad \forall t \in \mathbb{N}.$$

Exercise 2.20 (*Speed in unbalanced sequence*) The goal of this exercise is to show that the rate of convergence of a Metropolis random walk goes to one on a sequence of graphs, if there is a node whose degree vanishes compared to the degree of its neighbors. Let there be a sequence of symmetric connected graphs of increasing size $G_n = (V_n, E_n)$ and a sequence of nodes $v_n \in V_n$ such that

$$\lim_{n \to +\infty} \frac{d_{v_n}}{\min\{d_w : w \in \mathcal{N}_{v_n}\}} = 0.$$

Let P_n be the Metropolis random walk associated with G_n and $\rho_2^{(n)}$ its second largest eigenvalue. Using Cheeger bound, show that the gap $1 - \rho_2^{(n)}$ goes to zero when n diverges.

Exercise 2.21 (*Matrices adapted to $K_{m,n}$*) Let α, β be real numbers, and let the $(m + n)$-dimensional square matrix $M^{(\alpha,\beta)}$ be

$$M^{(\alpha,\beta)} = \begin{bmatrix} n\alpha I_m & -\alpha \mathbf{1}_m \mathbf{1}_n^* \\ -\beta \mathbf{1}_n \mathbf{1}_m^* & m\beta I_n \end{bmatrix}.$$

Verify that $M^{(\alpha,\beta)}$ has eigenvalues

- 0 corresponding to eigenvector $\mathbf{1}_{m+n}$;
- $n\alpha + m\beta$ corresponding to eigenvector $\begin{bmatrix} n\alpha\mathbf{1}_m \\ -m\beta\mathbf{1}_n \end{bmatrix}$;
- $n\alpha$ corresponding to the $(m-1)$-dimensional eigenspace span $\{\begin{bmatrix} x \\ \mathbf{0}_n \end{bmatrix} : x^*\mathbf{1} = 0\}$;
- $m\beta$ corresponding to the $(n-1)$-dimensional eigenspace span $\{\begin{bmatrix} \mathbf{0}_m \\ y \end{bmatrix} : y^*\mathbf{1} = 0\}$.

Exercise 2.22 (*Matrices adapted to wheels*) Let A_n be a normal[2] matrix of order n such that $A\mathbf{1}_n = d\mathbf{1}_n$, and denote by $x^{(k)}$ and $\lambda^{(k)}$ for $k \in \{1, \dots, n-1\}$ the remaining (orthonormal) eigenvectors of A_n with the corresponding eigenvalues. Consider matrix

$$M = \begin{bmatrix} 0 & \frac{1}{n}\mathbf{1}_n^* \\ \frac{1}{d+1}\mathbf{1}_n & \frac{1}{d+1}A_n \end{bmatrix}.$$

Verify that matrix M is stochastic and has eigenvalues 1 (simple), $-\frac{1}{d+1}$ (with eigenvector $\begin{bmatrix} -(d+1) \\ \mathbf{1}_n \end{bmatrix}$), and $\frac{\lambda^{(k)}}{d+1}$ for $k \in \{1, \dots, n-1\}$ (with eigenvector $\begin{bmatrix} 0 \\ x^{(k)} \end{bmatrix}$).

Exercise 2.23 (*Speed and democracy on $K_{m,n}$*) Let A, B be two sets such that $|A| = m \le n = |B|$ and consider the complete bipartite graph $K_{m,n} = (A \cup B, E)$ as defined in Example 1.2. Define on this graph

- the lazy simple random walk P by $P_{vw} = \frac{1}{2}\frac{1}{d_v}$ for all $(v, w) \in E$; and
- the lazy Metropolis random walk $\bar{P}$ by $\bar{P}_{vw} = \frac{1}{2}\min\{\frac{1}{d_v}, \frac{1}{d_w}\}$ for all $(v, w) \in E$.

Let π and ρ_2 be the invariant probability measure and the second largest eigenvalue of P, and correspondingly, let $\bar{\pi}$ and $\bar{\rho}_2$ be the invariant probability measure and the second largest eigenvalue of $\bar{P}$.

(i) Using Exercise 2.21, prove that the LSRW P is such that $\rho_2 = \frac{1}{2}$ and the invariant measure π is such that if $a \in A$, then $\pi_a = \frac{1}{2m}$, and if $b \in B$, then $\pi_b = \frac{1}{2n}$.
(ii) Prove that the LMRW $\bar{P}$ is such that $\bar{\pi}_v = \frac{1}{m+n}$ for all $v \in (A \cup B)$, and $\bar{\rho}_2 = 1 - \frac{m}{2n}$.
(iii) Let $m \in \mathbb{N}$ be fixed and $\{K_{m,n}\}_{n \ge m}$ a sequence of complete bipartite graphs of increasing size. Consider the consensus algorithms associated with these graphs by the above two definitions of random walks and compare them. Observe that the choice of either definition of the adapted stochastic matrix implies a trade-off between democracy and good convergence speed.

[2] A matrix A is said to be *normal* when $A^*A = AA^*$. Normal matrices are precisely those for which a complete basis of eigenvectors exists. Symmetric matrices are normal.

Exercise 2.24 (*Speed and democracy on wheels*) Let $n \geq 3$ and consider the n wheel graph W_n, which is defined as the union graph of a cycle C_n having node set $\{1, \ldots, n\}$ and a star S_n having node set $\{0\} \cup \{1, \ldots, n\}$. Define on this graph

- the lazy simple random walk P by setting $P_{vw} = \frac{1}{2}\frac{1}{d_v}$ for all $(v, w) \in E$; and
- the lazy Metropolis walk $\bar{P}$ by setting $\bar{P}_{vw} = \frac{1}{2}\min\{\frac{1}{d_v}, \frac{1}{d_w}\}$ for all $(v, w) \in E$.

Let π and ρ_2 be the invariant probability measure and the second largest eigenvalue of P and correspondingly let $\bar{\pi}$ and $\bar{\rho}_2$ be the invariant probability measure and the second largest eigenvalue of $\bar{P}$.

(i) Prove that the LSRW P is such that $\rho_2 \leq \frac{5}{6}$, and the invariant measure is such that $\pi_0 = \frac{1}{4}$ and $\pi_v = \frac{3}{4}\frac{1}{n}$ if $v \neq 0$.
(ii) Prove that the LMRW $\bar{P}$, which has uniform invariant measure, is such that $1 - \bar{\rho}_2 \leq \frac{7}{3}\frac{1}{n}$.
(iii) Now consider a sequence of wheel graphs of increasing size $\{W_n\}_{n \geq 3}$. Consider the consensus algorithms associated with these graphs by the above definitions of random walks. Remark that the choice of either definition of the adapted stochastic matrix implies a choice between democracy and good convergence speed.

Additional Topics

Exercise 2.25 (*Properties of Laplacians*) Let $G = (V, E, A)$ be a weighted graph of order n and L be the (weighted) Laplacian of G. Then, $\mathrm{rank}(L) = n - 1$ if and only if G contains a globally reachable vertex.

Exercise 2.26 (*Consensus in continuous time*) Consider a graph $G = (V, E)$ whose nodes are equipped with scalar dynamical systems $\dot{x}_v = u_v$, where $u \in \mathbb{R}^V$ is a control to be designed in order to achieve consensus. Let $A \in \mathbb{R}^{V \times V}$ be any nonnegative matrix such that $G = G_A$. Consider the feedback control law

$$u = -L(A)x.$$

(i) Verify that the control law $u = -L(A)x$ can be written componentwise as

$$u_v = \sum_w A_{vw}(x_w - x_v) \qquad \forall v \in V.$$

Consequently, it may be implemented by communicating with neighbors only and exchanging only relative information.

(ii) Show that, provided G has a globally reachable node, the closed loop system

$$\dot{x} = -L(A)x \tag{2.14}$$

yields consensus: For every initial condition $x(0)$, there exists a consensus value $\bar{x} \in \mathbb{R}$ such that $x_v(t) \to \bar{x}$ when $t \to +\infty$ for every $v \in V$. *Hint:* use Exercise 2.25.

(iii) Find the rate of convergence of (2.14).

Exercise 2.27 (*Node counting on a tree by message-passing*) Message-passing is a powerful approach to distributed computation, at least when the graph is a tree. Suppose then $G = (V, E)$ to be a tree and consider the following algorithm for the distributed computation of the number vertices N. Each unit $v \in V$ keeps in memory $d_v + 1$ scalar numbers, where d_v is the degree of v. We denote them as z_v^w with $w \in \mathcal{N}_v \cup \{v\}$. The algorithm is based on sending messages and updating z_v^w, according to the following rules.

- (Initialization): Set $z_v^w = 1$ for all $(v, w) \in E$.
- (Condition to send a message): Once unit v has received a message from all its neighbors except w, then v sends to w the following message: $z_{(v,w)} = z_v^w$.
- (Update upon receiving a message): When a node v receives $z_{(u,v)}$, the node does the following: $z_v^w = z_v^w + z_{(u,v)}$ for all $w \neq u$.
- (Termination): Once unit v has received message from all its neighbors, and updated z_w^v accordingly, node v sends messages $z_{(v,w)} = z_v^w$ to all neighbors w to whom no message has been send from v yet.

Verify that

(i) upon initialization, there is at least a node which satisfies the "condition to send a message";
(ii) on every directed edge (v, w), the message $z_{(v,w)}$ is transmitted exactly once, so that $2N - 2$ messages are exchanged over the network in total;
(iii) the algorithm terminates in finite time;
(iv) upon termination, $z_v^v = N$ for all $v \in V$.

Exercise 2.28 (*Consensus on a tree by message-passing*) Message-passing is a powerful approach to distributed computation, at least when G is a tree.

(i) Show that the procedure in Exercise 2.27—with a suitable initialization—may be used to compute the sum of N numbers given at the nodes, $x_v \in \mathbb{R}$ for $v \in V$.
(ii) Design a message-passing algorithm to compute the average of the x_v's.

Exercise 2.29 (*De Bruijn graphs and consensus* [14]) Consider De Bruijn graphs B_h^k on k symbols of dimension h.

(i) Design an algorithm, adapted to a De Bruijn graph, which converges in finite time to consensus.
(ii) How many steps does it take? Compare this value with the diameter and the degree of B_h^k.

Bibliographical Notes

Consensus problems have a very long history in social sciences [8, 13, 18], in statistics [10], and in computer science [40]. Their appearance in the field of control dates back to the thesis work of Tsitsiklis [49, 50] in the 80's, before a surge of interest about fifteen years ago, sparkled by the works [16, 28, 38]. Since then, the literature on the topic has grown enormously, motivated by the broad range of applications: rendezvous, deployment and formation control in robotic coordination [17], flocking of natural and artificial groups [52], load balancing in networks of processors and queues [12], clock synchronization [7, 32], optimal resource allocation [54], distributed optimization [37], distributed computation [24], distributed estimation and learning in sensor networks [44], social network analysis [27], and synchronization of interconnected systems [46, 53]. The averaging dynamics is ubiquitous to these problems (and many others): It is thus unsurprising that several books deal with the topic [5, 6, 31, 33, 41, 42].

This chapter gives a self-contained and comprehensive analysis of the "standard consensus algorithm" on time-invariant networks. In most prior works, its convergence properties are derived from the general theory of nonnegative matrices and in particular as corollaries of the Perron–Frobenius theorem (cf. [19, 45] for two classical references). This choice has two drawbacks. First, it is unnecessary because the needed results can be derived directly in an intuitive way. Second, since Perron–Frobenius theory does not extend to time-dependent networks, it hides the intimate connection between time-invariant and time-varying settings. Instead, the results from this chapter will be the foundation for the rest of the book. The main convergence principle in Lemma 2.1 is based on the presentation in Hendrickx's thesis [22, Sect. 9.2.1]. This principle is crucial and in this chapter we derived from it several properties of stochastic matrices that are central to our theory.

In probability theory, an important reason for the interest in stochastic matrices is the notion of Markov chain associated with it. While we refrain from introducing Markov chains in this text, we believe that the probabilistic interpretation of stochastic matrices is very useful. For instance, it motivates our discussion about the vector π. For these reasons, material on the theory of Markov chains can be a helpful additional reading: We recommend the textbook [30] and the monograph [1] that concentrates on reversible chains. More generally, the general theory of nonnegative matrices is an important background of our work.

Our analysis has highlighted the role of the second eigenvalue of the update matrix $\rho_2(P)$, which determines the speed of convergence of the average dynamics. In turn, the second eigenvalue is closely related to the spectral gap λ_2 of the associated graph. For this reason, graphs with a large spectral gap are of special interest to us. Graphs with large spectral gap are called *spectral expanders* and have been extensively studied in the last decades [2, 3, 43]. A serious study of expander graphs is outside the scope of this book. We only recall that De Bruijn graphs [55], defined in Chap. 1, have good expansion properties that, indeed, have been exploited in consensus problems [14].

In Sect. 2.6, we have shown how to construct stochastic matrices for a given topology. These constructions include the simple random walk and the Metropolis random walk, which are important examples all along the book. Actually, designing stochastic matrices according to certain performance criteria gives rise to a wide family of very interesting problems. For instance, one can look for doubly stochastic matrices, as we did specifically in Proposition 2.6: Paper [20] provides distributed algorithms to solve this design problem. Democracy is a milder requirement that is actually robust to perturbations of the graph, as shown in [11, 15].

Otherwise, one can seek to optimize the speed, that is, minimize the second eigenvalue. This problem is equivalent to finding the fastest mixing Markov chain and has been extensively studied, showing it to be a convex optimization problem [4]. Other (convex) performance metrics will be defined in Chap. 4. A related (more academic) question is finding the slowest possible convergence rate. For SRW matrices, the slowest rate is $1 - \gamma n^{-3}$, as proved in [29].

In a effort to overcome these slow convergence rates (recall also Exercise 2.12), some researchers have designed other distributed algorithms that guarantee a certain convergence speed, irrespective of the graph topology. For instance, the algorithm in [39] has a guaranteed $1 - cn^{-1}$ rate: We refer the reader to that paper also for several pointers to other "accelerated" consensus algorithms.

Our presentation of averaging dynamics has left aside a few topics which have attracted the interest of researchers and which we admit to be important: An incomplete list includes (i) consensus algorithms converging in finite time, (ii) consensus algorithms based on the "message-passing" approach, and (iii) consensus systems evolving in a continuous-time domain. We briefly discuss these natural issues here.

(i) The consensus algorithms presented in this chapter converge to consensus *asymptotically*. One can instead be interested in designing algorithms which converge in a finite number of step (necessarily, not smaller than the diameter). A trivial example is Example 2.6 for the complete graph, while other relatively simple examples can be constructed on De Bruijn graphs, see Exercise 2.29 and [14]. Actually, finite-time convergence can be obtained in more general graphs if we allow the update matrix to change with time (as we shall do in Chap. 3): A simple example are hypercubes [12, Sect. 4], but more general constructions are possible, see [23, 25, 47, 48].

(ii) Message-passing is a paradigm for distributed computation over networks, which we present through simple instances in Exercises 2.27 and 2.28. Nodes are thought of as objects with computational capabilities which can receive messages from their neighbors, elaborate them, and transmit them further. See, e.g., [34] for a general reference and [35] for an application to consensus. Message-passing has also recently found application in problems of leader selection, which will be defined in Chap. 5 [51].

(iii) In our work, we focus on discrete-time dynamics. However, much literature is concerned with continuous-time systems. For the time-invariant networks considered in this chapter, the analysis for continuous time and discrete time is closely related: Actually, the main results about the former can be derived as corollaries of our analysis, see Exercise 2.26. Instead, the analysis in continuous time can become trickier when the network is time-varying (see Chap. 3) or the interactions between nodes are nonlinear. We do not try to survey all the differences here: A few possible references, besides the books mentioned above, are [9, 26, 36].

References

1. Aldous, D., Fill, J.A.: Reversible Markov Chains and Random Walks on Graphs (2002). Unfinished monograph, recompiled 2014
2. Alon, N.: Eigenvalues and expanders. Combinatorica **6**(2), 83–86 (1986)
3. Alon, N., Roichman, Y.: Random Cayley graphs and expanders. Random Struct. Algorithms **5**, 271–284 (1994)
4. Boyd, S., Diaconis, P., Xiao, L.: Fastest mixing Markov chain on a graph. SIAM Rev. **46**(4), 667–689 (2004)
5. Bullo, F.: Lectures on Network Systems. Version 0.95 (2017). With contributions by Cortés, J., Dörfler, F., Martínez, S
6. Bullo, F., Cortés, J., Martínez, S.: Distributed Control of Robotic Networks. Applied Mathematics Series. Princeton University Press (2009)
7. Carli, R., Chiuso, A., Schenato, L., Zampieri, S.: Optimal synchronization for networks of noisy double integrators. IEEE Trans. Autom. Control **56**(5), 1146–1152 (2011)
8. Cartwright, D., Harary, F.: Structural balance: a generalization of Heider's theory. Psychol. Rev. **63**(5), 277 (1956)
9. Ceragioli, F., De Persis, C., Frasca, P.: Discontinuities and hysteresis in quantized average consensus. Automatica **47**(9), 1916–1928 (2011)
10. Chatterjee, S., Seneta, E.: Towards consensus: some convergence theorems on repeated averaging. J. Appl. Probab. **14**(1), 89–97 (1977)
11. Como, G., Fagnani, F.: Robustness of large-scale stochastic matrices to localized perturbations. IEEE Trans. Network Sci. Eng. **2**(2), 53–64 (2015)
12. Cybenko, G.: Dynamic load balancing for distributed memory multiprocessors. J. Parallel Distrib. Comput. **7**(2), 279–301 (1989)
13. DeGroot, M.H.: Reaching a consensus. J. Am. Stat. Assoc. **69**(345), 118–121 (1974)
14. Delvenne, J.-C., Carli, R., Zampieri, S.: Optimal strategies in the average consensus problem. Syst. Control Lett. **58**(10–11), 759–765 (2009)
15. Fagnani, F., Delvenne, J.-C.: The robustness of democratic consensus. Automatica **52**, 232–241 (2015)
16. Fax, J.A., Murray, R.M.: Information flow and cooperative control of vehicle formations. IEEE Trans. Autom. Control **49**(9), 1465–1476 (2004)
17. Francis, B.A., Maggiore, M.: Flocking and Rendezvous in Distributed Robotics. Springer (2015)
18. French, J.R.P.: A formal theory of social power. Psychol. Rev. **63**, 181–94 (1956)
19. Gantmacher, F.R.: The theory of matrices. Chelsea, New York (1959)
20. Gharesifard, B., Cortés, J.: Distributed strategies for generating weight-balanced and doubly stochastic digraphs. Eur. J. Control **18**(6), 539–557 (2012)
21. Golub, B., Jackson, M.O.: Naïve learning in social networks and the hegsel of crowds. Am. Econ. J. Microecon. **2**(1), 112–149 (2010)
22. Hendrickx, J.M.: Graphs and Networks for the Analysis of Autonomous Agent Systems. Ph.D. thesis, Université catholique de Louvain (2008)
23. Hendrickx, J.M., Jungers, R.M., Olshevsky, A., Vankeerberghen, G.: Graph diameter, eigenvalues, and minimum-time consensus. Automatica **50**(2), 635–640 (2014)
24. Hendrickx, J.M., Olshevsky, A., Tsitsiklis, J.N.: Distributed anonymous discrete function computation. IEEE Trans. Autom. Control **56**(10), 2276–2289 (2011)
25. Hendrickx, J.M., Shi, G., Johansson, K.H.: Finite-time consensus using stochastic matrices with positive diagonals. IEEE Trans. Autom. Control **60**(4), 1070–1073 (2015)
26. Hendrickx, J.M., Tsitsiklis, J.N.: Convergence of type-symmetric and cut-balanced consensus seeking systems. IEEE Trans. Autom. Control **58**(1), 214–218 (2013)

27. Jackson, M.O.: Social and Economic Networks. Princeton University Press (2010)
28. Jadbabaic, A., Lin, J., Morse, A.S.: Coordination of groups of mobile autonomous agents using nearest neighbor rules. IEEE Trans. Autom. Control **48**(6), 988–1001 (2003)
29. Landau, H.J., Odlyzko, A.M.: Bounds for eigenvalues of certain stochastic matrices. Linear Algebra Appl. **38**, 5–15 (1981)
30. Levin, D.A., Peres, Y., Wilmer, E.L.: Markov Chains and Mixing Times. American Mathematical Society (2009)
31. Lewis, F.L., Zhang, H., Hengster-Movric, K., Das, A.: Cooperative Control of Multi-agent Systems: Optimal and Adaptive Design Approaches. Springer Science & Business Media (2013)
32. Li, Q., Rus, D.: Global clock syncronization in sensor networks. IEEE Trans. Comput. **55**(2), 214–226 (2006)
33. Mesbahi, M., Egerstedt, M.: Graph Theoretic Methods for Multiagent Networks. Applied Mathematics Series. Princeton University Press, 0 (2010)
34. Mézard, M., Montanari, A.: Information, Physics, and Computation. Oxford University Press (2009)
35. Moallemi, C.C., Van Roy, B.: Consensus propagation. IEEE Trans. Inf. Theory **52**(11), 4753–4766 (2006)
36. Moreau, L.: Stability of continuous-time distributed consensus algorithms. In: IEEE Conference on Decision and Control, Bahamas, pp. 3999–4003, December 2004
37. Nedich, A.: Convergence rate of distributed averaging dynamics and optimization in networks. Found. Trends® Syst. Control, **2**(1), 1–100 (2015)
38. Olfati-Saber, R., Murray, R.M.: Consensus problems in networks of agents with switching topology and time-delays. IEEE Trans. Autom. Control **49**(9), 1520–1533 (2004)
39. Olshevsky, A.: Linear time average consensus on fixed graphs and implications for decentralized optimization and multi-agent control, May 2016
40. Pease, M., Shostak, R., Lamport, L.: Reaching agreement in the presence of faults. J. Assoc. Comput. Mach. **27**(2), 228–234 (1980)
41. Ren, W., Beard, R.W.: Distributed Consensus in Multi-vehicle Cooperative Control Theory and Applications. Springer, New York (2007)
42. Ren, W., Cao, Y.: Distributed Coordination of Multi-agent Networks: Emergent Problems, Models, and Issues. Springer Science & Business Media (2011)
43. Sarnak, P.: What is...an expander? Not. Am. Math. Soc. **51**(7), 762–763 (2004)
44. Sayed, A.H.: Adaptation, learning, and optimization over networks. Found. Trends® Mach. Learn. **7**(4–5), 311–801 (2014)
45. Seneta, E.: Non-negative Matrices and Markov Chains. Springer Series in Statistics. Springer, Berlin (2006)
46. Strogatz, S.H.: From Kuramoto to Crawford: exploring the onset of synchronization in populations of coupled oscillators. Physica D **143**(1–4), 1–20 (2000)
47. Sundaram, S., Hadjicostis, C.N.: Finite-time distributed consensus in graphs with time-invariant topologies. In: American Control Conference, pp. 711–716, July 2007
48. Tran, T.D., Kibangou, A.Y.: Distributed design of finite-time average consensus protocols. In: IFAC Workshop on Estimation and Control of Networked Systems, Koblenz, Germany, pp. 227–233, September 2013
49. Tsitsiklis, J.N.: Problems in Decentralized Decision Making and Computation. Ph.D. thesis, Massachusetts Institute of Technology, November 1984
50. Tsitsiklis, J.N., Bertsekas, D.P., Athans, M.: Distributed asynchronous deterministic and stochastic gradient optimization algorithms. IEEE Trans. Autom. Control **31**(9), 803–812 (1986)
51. Vassio, L., Fagnani, F., Frasca, P., Ozdaglar, A.: Message passing optimization of harmonic influence centrality. IEEE Trans. Control Network Syst. **1**(1), 109–120 (2014)

52. Vicsek, T., Czirók, A., Ben-Jacob, E., Cohen, I., Shochet, O.: Novel type of phase transition in a system of self-driven particles. Phys. Rev. Lett. **75**(6), 1226–1229 (1995)
53. Wieland, P., Sepulchre, R., Allgöwer, F.: An internal model principle is necessary and sufficient for linear output synchronization. Automatica **47**(5), 1068–1074 (2011)
54. Xiao, L., Boyd, S.: Optimal scaling of a gradient method for distributed resource allocation. J. Optim. Theory Appl. **129**(3), 469–488 (2006)
55. Zhang, F.J., Lin, G.N.: On the de Bruijn-Good graphs. Acta Math. Sinica **30**(2), 195–205 (1987)

Chapter 3
Averaging in Time-Varying Networks

Abstract This chapter studies averaging dynamics in which the update matrix and, possibly, the underlying graph may be different at each time step. This extension is particularly important for the applications. Indeed, in realistic models of sensor and robotic networks, units, and links may be occasionally off due to environmental reasons or for energy saving purposes. Similarly, social dynamics may involve complex patterns of interactions that change over time. We are going to show that time-dependent consensus algorithms converge under relatively mild assumptions involving suitable notions of connectivity. Actually, the underlying graph needs not to be connected at any time, but the sequence of graphs must be "sufficiently connected" over time. More specifically, in Sects. 3.1 and 3.2 we provide two families of results, corresponding to two sufficient connectivity assumptions. Our presentation also includes, in Sect. 3.3, cases when the matrix evolves randomly in time. These randomized dynamics encompass the so-called *gossip* algorithms, which have attracted much attention in the last decade.

3.1 Time-Varying Updates: Uniform Connectivity

Given a set of nodes V of cardinality N, we consider a distributed state $x(t) \in \mathbb{R}^V$ evolving according to a system of the form

$$x(t+1) = P(t)x(t) \qquad t \in \mathbb{Z}_{\geq 0}, \tag{3.1}$$

where $P(t)$ is a stochastic matrix for each $t \geq 0$. We will use the following notation

$$P(s,s) = I, \quad P(t,s) = P(t-1)\dots P(s), \quad 0 \leq s < t$$

so that $x(t) = P(t,s)x(s)$ for every $s \leq t$.

We start with a preliminary result which is a simple consequence of the contraction principle Lemma 2.1 already used in the time-invariant context.

© Springer International Publishing AG 2018

F. Fagnani and P. Frasca, *Introduction to Averaging Dynamics over Networks*, Lecture Notes in Control and Information Sciences 472,
https://doi.org/10.1007/978-3-319-68022-4_3

Lemma 3.1 *Consider system* (3.1). *Assume that*

(*i*) *there exists* $\alpha \in (0, 1]$ *such that, for every* $t \geq 0$ *and* $u, v \in V$, $P_{uv}(t) > 0$ *implies* $P_{uv}(t) \geq \alpha$;

(*ii*) *there exists a sequence of times* $\{t_k \in \mathbb{Z}_{\geq 0} : k \in \mathbb{Z}_{\geq 0}\}$ *such that*

  (*a*) *there exists* $B \in \mathbb{N}$ *such that* $t_{k+1} - t_k \leq B$ *for all* k *and*
  (*b*) *for every* k, *there exists* $v^* \in V$ *such that,* $P(t_{k+1}, t_k)_{uv^*} > 0$ *for every* $u \in V$.

Then, $x(t)$ *converges to a point in* span$\{\mathbf{1}\}$ *from every initial condition in* $\mathbb{R}^V$.

Proof Notice first of all that thanks to assumptions (i) and (ii), we have that

$$P(t_{k+1}, t_k)_{uv^*} > \alpha^{t_{k+1}-t_k} \geq \alpha^B, \quad \forall u \in V. \tag{3.2}$$

Define now, for every $t \geq 0$,

$$x_{\min}(t) = \min_u \{x_u(t)\} \qquad x_{\max}(t) = \max_u \{x_u(t)\}$$

and notice that Lemma 2.1 together with (3.2) implies that, for every $k \geq 0$,

$$x_{\max}(t_{k+1}) - x_{\min}(t_{k+1}) \leq (1 - \alpha^B)\big(x_{\max}(t_k) - x_{\min}(t_k)\big) \tag{3.3}$$

Considering that $x_{\max}(t)$ and $x_{\min}(t)$ are two monotonic sequences, thus admitting limit, it follows that $x_{\max}(t) - x_{\min}(t) \to 0$ as $t \to +\infty$. This yields the thesis. □

The above result is not very appealing for application as condition (ii). needs in principle to consider large products of the matrices $P(t)$ determining the dynamics. As in the time-invariant case, we would like to have results whose assumptions are at the level of the associated graphs $\mathcal{G}_{P(t)}$. The reader may recall that, on static networks, consensus was proved (see Theorem 2.2) under two conditions: a connectivity condition and an aperiodicity condition. On time-varying networks, we are going to make suitable assumptions on the connectivity over time, while the aperiodicity condition is replaced by the following assumption.

Definition 3.1 (*Nondegeneracy*) A set $\mathcal{P}$ of stochastic matrices over V is *nondegenerate* if

(i) for every $P \in \mathcal{P}$ and for every $u \in V$, $P_{uu}(t) > 0$;
(ii) there exists $\alpha \in (0, 1]$ such that, for every $P \in \mathcal{P}$ and $u, v \in V$, $P_{uv}(t) > 0$ implies $P_{uv}(t) \geq \alpha$.

It is clear that the assumption of nondegeneracy relates to aperiodicity: Indeed, if $\mathcal{P}$ is nondegenerate, then each $P \in \mathcal{P}$ is aperiodic. Notice that the converse is not true, because of the strong positivity condition expressed in the definition. Moreover, the mere aperiodicity of each matrix $P(t)$ is not sufficient for consensus; see Exercise 3.1.

The following result shows some fundamental consequences of nondegeneracy. If P is a stochastic matrix over V, below we will use the notation $\mathcal{G}_P = (V, E_P)$ for the graph associated with P.

Proposition 3.1 *Suppose that $P(t)$, for $t \in \mathbb{Z}_{\geq 0}$ is a non degenerate sequence of stochastic matrices. Fix $t_1 \leq t_2 \leq t_3 \leq t_4$. Then,*

(i) $E_{P(t_3,t_2)} \subseteq E_{P(t_4,t_1)}$*;*
(ii) If $(u, v) \in E_{P(t_4,t_3)}$ *and* $(v, w) \in E_{P(t_2,t_1)}$*, then* $(u, w) \in E_{P(t_4,t_1)}$*.*

Proof Both claims follow immediately by combining the two inequalities

$$P(t_4, t_1)_{uw} \geq P(t_4, t_3)_{uu} P(t_3, t_2)_{uw} P(t_2, t_1)_{ww}$$
$$P(t_4, t_1)_{uw} \geq P(t_4, t_3)_{uv} P(t_3, t_2)_{vv} P(t_2, t_1)_{vw}$$

and property (i) of nondegeneracy. □

We are now ready to state the main convergence result of this section which is a generalization of Theorem 2.2.

Theorem 3.1 (Time-dependent consensus I) *Consider system* (3.1). *Assume that*

(i) the set of matrices $\{P(t)\}$ is nondegenerate;
(ii) there exists a duration $T \in \mathbb{N}$ such that, for all $t_0 \in \mathbb{Z}_{\geq 0}$, the graph

$$\bigcup_{s=0}^{T-1} \mathcal{G}_{P(t_0+s)}$$

contains a globally reachable node.

Then, $x(t)$ converges to a point in $\text{span}\{\mathbf{1}\}$ *from every initial condition in* $\mathbb{R}^V$.

The reader should note that the connectivity condition does not imply anything on each single graph $\mathcal{G}_{P(t)}$, which may well be not connected at any time t. The following example illustrates the application of the theorem.

Example 3.1 (*Sequences of graphs*) Consider the following sequences composed of the graphs represented in Fig. 3.1.

$$\text{(i)}\ \ \mathcal{S}_1(t) = \begin{cases} G_a & \text{if } t \text{ is a square number} \\ G_b & \text{otherwise} \end{cases}$$

$$\text{(ii)}\ \ \mathcal{S}_2(t) = \begin{cases} G_b & \text{if } t \text{ is a square number} \\ G_a & \text{otherwise} \end{cases}$$

$$\text{(iii)}\ \ \mathcal{S}_3(t) = \begin{cases} G_c & \text{if } t \text{ is a square number} \\ G_d & \text{otherwise} \end{cases}$$

$$\text{(iv)}\ \ \mathcal{S}_4(t) = \begin{cases} G_d & \text{if } t \text{ is a square number} \\ G_a & \text{otherwise} \end{cases}$$

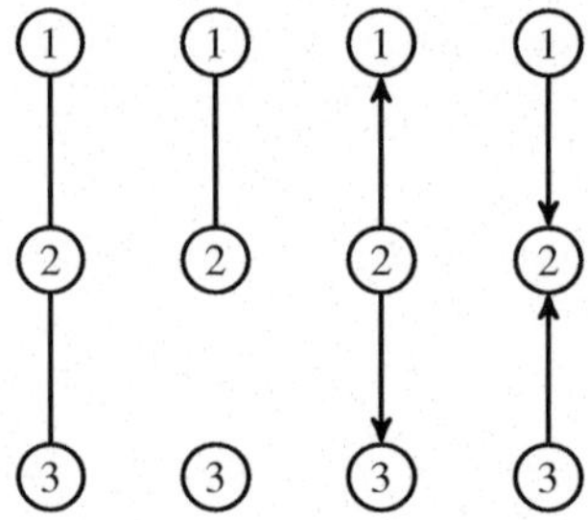

Fig. 3.1 The graphs G_a, G_b, G_c, and G_d used in Example 3.1

Let now $P_i(t)$ denote the sequence of SRW matrices constructed on the sequence of graphs $\mathscr{S}_i(t)$. By Theorem 3.1, we conclude that $P_2(t)$ and $P_3(t)$ lead to consensus. Instead, nothing can be concluded regarding $P_1(t)$ or $P_4(t)$ because assumption (ii) in Theorem 3.1 is not satisfied. Both cases actually lead to a consensus. For $P_4(t)$, consensus is trivial by observing that the SRW related to G_d leads to consensus in one step, while for $P_1(t)$ consensus will follow by Corollary 3.2 later on.

The proof of Theorem 3.1 relies on the results proven so far as well on a classical combinatorial argument reported below for the convenience of the reader.

Lemma 3.2 (Pigeonhole principle) *If n discrete objects are to be allocated to m containers, then at least one container must hold no fewer than $\lceil \frac{n}{m} \rceil$ objects.*

Proof (of Theorem 3.1) Notice first of all that since

$$\bigcup_{s=0}^{T-1} \mathscr{G}_{P(t_0+s)} \subseteq \mathscr{G}_{P(t_0+T,t_0)}$$

by virtue of Proposition 3.1, it follows that the graph $\mathscr{G}_{P(t_0+T,t_0)}$ contains a globally reachable node for every t_0. Let $N = |V|$ and let M be the number of distinct graphs over V possessing a globally reachable node. Consider the time interval $[0, NMT[$ split into subintervals $[0, T[$, $[T, 2T[$ and so on. By the pigeonhole principle, there must exist a graph G over V possessing a globally reachable node, which is repeated at least N times among the sequence of graphs $\mathscr{G}_{P(jT,(j-1)T)}$ for $j = 0, \ldots, NM$. Denote by v^* the globally reachable node inside G. Since every node $u \in V$ is connected to v^* in G with a path of length $l \leq N$, it follows by a repeated application of Proposition 3.1 that $\mathscr{G}_{P(NMT,0)}$ contains every edge of type (u, v^*) for $u \in V$. This implies that $P(NMT, 0)_{uv^*} > 0$ for every $u \in V$. Arguing similarly on every matrix $P(kNMT, (k-1)NMT)$, we can see that the assumptions of Lemma 3.1 are satisfied if we take $t_k = kNMT$. □

Remark 3.1 (*Convergence time*) Note that the construction in the proof of Theorem 3.1 is "worst-case" in nature and gives little clue about the actual convergence time for the algorithm. This issue is investigated in an exemplary case in Exercise 3.2.

Remark 3.2 (*Consensus point*) In general, the final value upon which all the states x_u agree in the limit is unknown. This final value depends on the initial condition and the specific sequence of matrices defining the time-dependent linear algorithm. Only in few cases, one can compute the final value. One instance are time-independent consensus algorithms, which have been considered in the previous chapter. Another instance are time-dependent algorithms (3.1) involving doubly stochastic matrices. Indeed, it is clear that whenever $P(t)$ is doubly stochastic for every $t \geq 0$, then $x_{\text{ave}}(t) = x_{\text{ave}}(0)$. Then, provided $x(t)$ converges, it converges to $x_{\text{ave}}(0)\mathbf{1}$. This simple remark can be immediately extended to any sequence of matrices which share their dominant left eigenvector.

Theorem 3.1 requires a *uniform* connectivity assumption: The union of graphs, over time, must be connected within a fixed window. Later on, we will present results where the connectivity assumption is weaker. The following result shows that some connectivity condition for consensus will, however, be necessary.

Proposition 3.2 (Connectivity is necessary) *Consider system* (3.1). *If, for every initial condition* $x(0)$, *the state* $x(t)$ *converges to a point in* $\text{span}\{\mathbf{1}\}$, *then there exists a node which is globally reachable in the graph*

$$G = \bigcup_{s \geq 0} \mathcal{G}_{P(s)}.$$

Proof By contradiction, assume that G does not possess a globally reachable node. This implies that the correspondent condensation graph has two leaves that correspond to two strongly connected subgraphs of G, denoted by $G_i = (V_i, E_i)$ for $i = 1, 2$ and such that $V_1 \cap V_2 = \emptyset$ and there is no path from V_1 to V_2 or from V_2 to V_1. Consider now an initial condition $x(0)$ such that $x(0)_v = 0$ for all $v \in V_1$ and $x(0)_v = 1$ for all $v \notin V_1$. Since there is no edge outgoing G_1 and G_2, clearly, if Q is a stochastic matrix adapted to G, it follows that $(Qx(0))_v = 0$ for all $v \in V_1$ and $Qx(0)_v = 1$ for all $v \in V_2$. This implies that $x(t)_v = 0$ for all $v \in V_1$ and $x(t)_v = 1$ for all $v \in V_2$, and thus, $x(t)$ cannot converge to a consensus. □

As the necessary condition in Proposition 3.2 is weaker than the sufficient condition in Theorem 3.1, it is natural to ask whether the former is sufficient as well. The answer is negative, as shown by the example proposed in Exercise 3.3.

3.2 Time-Varying Updates: Cut-Balanced Interactions

Theorem 3.1 requires a *uniform* connectivity assumption: The union of graphs, over time, must be connected within a fixed window. In this section, we seek conditions under which this uniform connectivity requirement can be dropped, while maintaining convergence.

To this goal, we need to introduce two new concepts, *cut-balanced* graph and *limit graph*. A graph $G = (V, E)$ is said to be *cut-balanced* when for any nonempty proper subset $S \subset V$, there exist $v \in S$ and $w \notin S$ with $(v, w) \in E$ if and only if there exist $v' \notin S$ and $w' \in S$ with $(v', w') \in E$. Clearly, if a graph is symmetric or strongly connected, then it is also cut-balanced. More precisely, cut-balanced graphs allow for the following characterization.

Lemma 3.3 *The graph G is cut-balanced if and only if every weakly connected component of G is strongly connected.*

Proof Assume that the graph is cut-balanced and let $W \subseteq V$ be a weakly connected component of G. If W is not strongly connected, then there exist a node $u \in W$ from which not all nodes in W can be reached. Let S be the subset of nodes which are reachable from u. Clearly, S is a proper subset of W and, necessarily, there is no edge from S to $W \setminus S$, while there must be edges from $W \setminus S$ to S; otherwise, W would not be a weakly connected component. This clearly contradicts the fact that G was cut-balanced. The proof of the reverse implication is left to the reader. □

The second key ingredient is the definition of *limit graph*. Given a sequence of graphs $(G_t)_{t \in \mathbb{N}}$ such that $G_t = (V, E_t)$ for all $t \in \mathbb{N}$, we say that the limit graph of this sequence is the graph $G_\infty = (V, E_\infty)$ where the edge set E_∞ equals to the set-theoretic limit superior of the sequence $(E_t)_t$, that is,

$$E_\infty = \limsup_{t \in \mathbb{N}} E_t := \bigcap_{t \geq 0} \bigcup_{s \geq 0} E_{t+s}.$$

Equivalently, an edge (u, v) is in E_∞ when $(u, v) \in E_n$ for infinitely many n. This limit graph "forgets" transient interactions and focuses on those interactions that occur infinitely often and thus affect the convergence behavior.

We are now ready to state and prove the main result of this section.

Theorem 3.2 (Convergence) *Consider system* (3.1). *Assume that*

(i) the set of matrices $\{P(t)\}$ is nondegenerate;
(ii) the associated graph $\mathcal{G}_{P(t)}$ is cut-balanced for every $t \geq 0$.

Then, $x(t)$ converges to a limit point $\tilde{x} \in \mathbb{R}^V$ such that $\tilde{x}_u \in [x_{\min}(0), x_{\max}(0)]$ for all $u \in V$. Furthermore, let G_∞ be the limit graph of the sequence $(\mathcal{G}_{P(t)})_t$. If two nodes v and w belong to the same connected component of G_∞, then $\tilde{x}_v = \tilde{x}_w$.

Proof Since $\mathcal{G}_{P(t)}$ is cut-balanced for every $t \geq 0$, then also G_∞ is cut-balanced: This implies, by Lemma 3.3, that all weakly connected components of G_∞ are strongly connected. Let $C \subset V$ denote the node set of one such connected component and observe that there exists a time $t_0 \geq 0$ such that $P_{vw}(t) = P_{wv}(t) = 0$ for all $v \in C$, $w \notin C$ and $t \geq t_0$. Then, without loss of generality we disregard the dynamics before t_0 and we study the dynamics for $t \geq t_0$ over the component C.

Let $m \in C$ be a node such that $x_m(t_0) = \max\{x_u(t_0) : u \in C\}$. Define $S_{t_0} = \{m\}$ and, iteratively for $t \geq t_0$, a sequence of subsets $S_t \subseteq V$, by

$$S_{t+1} = \{u \in C : \exists w \in S_t \text{ such that } P_{vw}(t) > 0\}.$$

The sequence of sets S_t collects those nodes whose states at time t are influenced by the state of node m at time t_0. Note that because of the nondegeneracy assumption, the inclusion $S_{t+1} \supseteq S_t$ holds for every $t \geq t_0$. Let $t^\star$ be the time at which S_t is maximal and assume by contradiction that $S_{t^\star} \neq C$. Then, there is no vertex outside $S_{t^\star}$ that is connected to any vertex in $S_{t^\star}$ for any time $t \geq t^\star$: So C is not strongly connected in the graph $\cup_{s \geq t^\star} \mathcal{G}_{P(s)}$, contradicting the assumptions. Hence, $S_{t^\star} = C$.

We claim that for every $t_0 \leq t \leq t^\star$ and every $v \in S_t$, it holds

$$x_v(t) \geq \min_{u \in C} x_u(t_0) + \alpha^{|S_t|-1}\big(\max_{u \in C} x_u(t_0) - \min_{u \in C} x_u(t_0)\big), \tag{3.4}$$

where $\alpha > 0$ is the nondegeneracy constant. This fact can be shown by induction on t. For $t = t_0$, we have $S_{t_0} = \{m\}$ and so (3.4) trivially holds. For the induction step, we need to consider two cases. If $S_{t+1} = S_t$, then at time t every $w \in S_t$ only influences nodes in S_t. By the cut-balance assumption, every $v \in S_t$ is then only influenced by nodes in S_t. Hence, for every $v \in S_{t+1}$

$$\begin{aligned} x_v(t+1) &= \sum_{w \in S_t} P_{vw}(t) x_w(t) \\ &\geq \sum_{w \in S_t} P_{vw}(t)\Big(\min_{u \in C} x_u(t_0) + \alpha^{|S_t|-1}\big(\max_{u \in C} x_u(t_0) - \min_{u \in C} x_u(t_0)\big)\Big) \\ &= \min_{u \in C} x_u(t_0) + \alpha^{|S_{t+1}|-1}\big(\max_{u \in C} x_u(t_0) - \min_{u \in C} x_u(t_0)\big). \end{aligned}$$

If instead $S_{t+1} \neq S_t$, we note that for every $v \in S_{t+1}$, there is at least one $w \in S_t$ and such that $P_{vw}(t) > 0$. Indeed, if $v \notin S_t$, then v is, by construction, connected to at least one node $w \in S_t$, whereas every $v \in S_t$ is by the hypothesis always connected to itself. Since all (positive) entries $P_{vw}(t)$ are by hypothesis lower-bounded by α, this together with the induction hypothesis implies that

$$\begin{aligned} x_v(t+1) &= \sum_{w \in V} P_{vw}(t) x_w(t) \\ &\geq \sum_{w \in S_t} P_{vw}(t)\Big(\min_{u \in C} x_u(t_0) + \alpha^{|S_t|-1}\big(\max_{u \in C} x_u(t_0) - \min_{u \in C} x_u(t_0)\big)\Big) + \sum_{w \notin S_t} P_{vw}(t) \min_{u \in C} x_u(t_0) \\ &\geq \min_{u \in C} x_u(t_0) + \alpha\, \alpha^{|S_t|-1}\big(\max_{u \in C} x_u(t_0) - \min_{u \in C} x_u(t_0)\big) \\ &\geq \min_{u \in C} x_u(t_0) + \alpha^{|S_{t+1}|-1}\big(\max_{u \in C} x_u(t_0) - \min_{u \in C} x_u(t_0)\big), \end{aligned}$$

thus proving (3.4). As $\max_{u \in C} x_u(t)$ is not increasing, inequality (3.4) implies that

$$\max_{u\in C} x_u(t^\star) - \min_{u\in C} x_u(t^\star) \leq (1-\alpha^{|C|-1})(\max_{u\in C} x_u(t_0) - \min_{u\in C} x_u(t_0)).$$

By repeating all the above reasoning starting from $t_1 = t^\star$ and so on, we can construct a sequence of times $\{t_k : k \geq 0\}$ such that for every k it holds that

$$\max_{u\in C} x_u(t_{k+1}) - \min_{u\in C} x_u(t_{k+1}) \leq (1-\alpha^{|C|-1})(\max_{u\in C} x_u(t_k) - \min_{u\in C} x_u(t_k)).$$

This fact implies that all nodes $u \in C$ converge to consensus and thus proves the result. □

Remarkably, Theorem 3.2 does not contain any connectivity assumption other than cut-balance. As a consequence, it does not guarantee consensus among all states, but only convergence and "local" consensus inside each connected component of the limit graph. Global consensus can instead be obtained by restoring an assumption of global connectivity, as in the following two results which immediately follow from Theorem 3.2.

Corollary 3.1 (Time-dependent consensus II) *Consider system* (3.1). *Assume that*

(i) the set of matrices $\{P(t)\}$ is nondegenerate;
(ii) for every $t \geq 0$, the graph $\mathcal{G}_{P(t)}$ is cut-balanced; *and*
(iii) for every $t \geq 0$, the graph $\bigcup_{s\geq 0} \mathcal{G}_{P(t+s)}$ is weakly connected.

Then, $x(t)$ converges to a point in span$\{\mathbf{1}\}$ *from every initial condition in $\mathbb{R}^V$.*

Note that condition (iii) is equivalent to G_∞ being strongly connected (as the reader may verify). As a special case, we recover the following result on convergence for symmetric graphs. We note that this theorem does not require the matrix $P(t)$ to be symmetric but just its induced graph to be symmetric, i.e., to encode *reciprocal* communications.

Corollary 3.2 (Time-dependent consensus III) *Consider system* (3.1). *Assume that*

(i) the set of matrices $\{P(t)\}$ is nondegenerate;
(ii) for every $t \geq 0$, the graph $\mathcal{G}_{P(t)}$ is symmetric; and
(iii) for every $t \geq 0$, the graph $\bigcup_{s\geq 0} \mathcal{G}_{P(t+s)}$ is connected.

Then, $x(t)$ converges to a point in span$\{\mathbf{1}\}$ *from every initial condition in $\mathbb{R}^V$.*

We stress that these two results do not require any uniform connectivity assumption, and indeed, their proofs do not rely on Lemma 2.1, as opposed to Theorem 3.1. As an example, Corollary 3.2 implies that the sequence $P_1(t)$ in Example 3.1 leads to a consensus.

The interest in a convergence result such as Theorem 3.2, which avoids connectivity assumptions, becomes more apparent in those contexts where checking connectivity is difficult, for instance because the evolution of $P(t)$ depends on the current state $x(t)$. We are going to illustrate this difficulty with a very popular example, known as *Krause's model*.

Example 3.2 (*Krause's model*) In this dynamics, agent v trusts, i.e., takes into account for its update, only those agents w whose current state $x_w(t)$ is close enough to $x_v(t)$. More precisely, we fix a threshold $\varepsilon > 0$ and, for all $t \in \mathbb{Z}_{\geq 0}$ and all $v \in V$, we let $N_v(t) = \{u \in V : |x_v(t) - x_u(t)| \leq \varepsilon\}$. Given $\rho \in (0, 1]$, we then define the dynamics

$$x_v(t+1) = x_v(t) + \frac{\rho}{|N_v(t)|} \sum_{w \in N_v(t)} (x_w(t) - x_v(t)) \qquad v \in V\,. \tag{3.5}$$

Convergence of dynamics (3.5) can be deduced from Theorem 3.2. Notice indeed that if we define

$$P_{vw}(t) = \begin{cases} 1 - \rho\frac{|N_v(t)|-1}{|N_v(t)|} & \text{if } w = v \\ \frac{\rho}{|N_v(t)|} & \text{if } w \in N_v(t),\, w \neq v \\ 0 & \text{if } w \notin N_v(t) \end{cases}$$

it is immediate to check that $x(t+1) = P(t)x(t)$. Notice that $\mathcal{G}_{P(t)}$ is symmetric so that assumption (ii) is verified. Nondegeneracy follows easily from the definition as $P_{vv}(t) \geq 1 - \rho\frac{N-1}{N}$ for every $v \in V$ while, if $v \neq w$ and $P_{vw}(t) > 0$, it follows that $P_{vw}(t) = \rho/|N_v(t)| \geq \rho/N$.

Notice that, in this model, the matrix $P(t)$ actually depends on the state of agents at time t and, as a consequence, the model is *nonlinear*. There is no way to guarantee a priori uniform connectivity conditions on the sequence of corresponding graphs, so

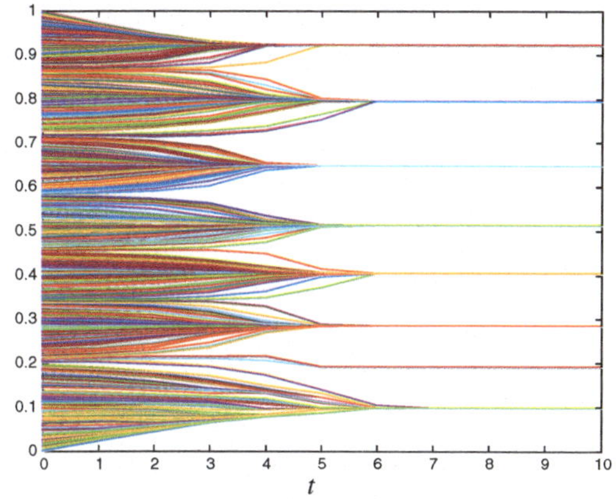

Fig. 3.2 A typical evolution under the time-varying dynamics (3.5) for a large number of agents: In this case, $N = 1000$ and $\varepsilon = 0.05$. Observe that clusters are approximately 2ε apart

that Theorem 3.1 cannot be applied. Simulations (see Fig. 3.2) demonstrate that the limit state is not a consensus point, but instead a collection of disconnected "clusters," composed of agents which share the same limit opinion. Moreover, if $\rho = 1$, then convergence is attained in finite time (see Exercise 3.4).

Krause's dynamics was originally proposed to model opinion dynamics with bounded confidence, but can also represent a simple model of one-dimensional vehicle rendezvous with limited visibility. This model has received much attention in the last years and many generalizations have been proposed: In Exercise 3.5, we study one of these.

3.3 Randomized Updates

This section presents time-varying consensus algorithms, where the update matrix is selected at each time step by a random process. Given a set of nodes V of finite cardinality N, we consider for every time $t \in \mathbb{Z}_{\geq 0}$ a random vector $x(t) \in \mathbb{R}^V$ evolving according to a random discrete-time system of the form

$$x(t+1) = P(t)x(t) \qquad t \in \mathbb{Z}_{\geq 0}, \tag{3.6}$$

where $P(t)$ is a stochastic matrix for each $t \geq 0$ and $\big(P(t)\big)_{t\geq 0}$ is a sequence of independent and identically distributed random variables. Note that the initial condition is unknown but fixed (not random) and that all the randomness originates from generating the sequence of $P(t)$s. Consequently, in what follows the phrase "almost surely" means "with probability 1" with respect to the matrix selection process.

We begin our discussion from the following example of randomized dynamics. Let a symmetric graph $G = (V, E)$ be given, and for each time step $t \geq 0$, let an edge (v, w) be chosen in E, according to a uniform distribution over E. Define

$$\begin{aligned}
x_v(t+1) &= \frac{1}{2}x_v(t) + \frac{1}{2}x_w(t)\,,\\
x_w(t+1) &= \frac{1}{2}x_w(t) + \frac{1}{2}x_v(t)\,,\\
x_u(t+1) &= x_u(t), \quad \text{for } u \neq v, w\,.
\end{aligned}$$

This dynamics can be written in the form (3.6) by defining

$$P^{(v,w)} = I - \frac{1}{2}(e_v e_v^* - e_v e_w^* - e_w e_v^* + e_w e_w^*),$$

where e_u is the uth vector of the canonical basis of $\mathbb{R}^V$, and $\mathbb{P}[P(t) = P^{(v,w)}] = \frac{1}{|E|}$.

We shall refer to this dynamics as the uniform symmetric gossip (USG).

Proposition 3.3 (USG convergence) *A uniform symmetric gossip dynamics converges almost surely to the average of the initial conditions, provided the underlying graph is connected.*

Proof For any $t_0 \geq 0$ and any edge $(v, w) \in E$, where E is the edge set of the underlying graph, we evaluate the probability of the event "the edge (v, w) is not selected for update at any time larger than t_0." Since the probability that (v, w) is not selected at any time t is $1 - \frac{1}{|E|}$, the probability that (v, w) is not selected for all times s such that $t_0 \leq s < t$ is $\left(1 - \frac{1}{|E|}\right)^{t-t_0}$. Since $\lim_{t \to +\infty} \left(1 - \frac{1}{|E|}\right)^{t-t_0} = 0$, we argue that (v, w) is selected infinitely often after t_0 with probability 1. This fact implies that $G_\infty = G$ almost surely. Since G is connected, convergence can be deduced from Corollary 3.2. □

In the following, we are going to present a more general convergence result that subsumes Proposition 3.3. We will rely on the results of Sect. 3.1. To begin with, let us go back to system (3.6) and let us study the *expected* dynamics, that is, the dynamics of $\mathbb{E}[x(t)]$. Equation (3.6) and the independence among $P(t)$s imply $\mathbb{E}[x(t+1)|x(t)] = \mathbb{E}[P(t)]x(t)$ for all t. Then, denoting $\bar{P} := \mathbb{E}[P(t)]$, we have

$$\mathbb{E}[x(t+1)] = \bar{P}\,\mathbb{E}[x(t)].$$

Note that $\bar{P}$ is a stochastic matrix. If the graph associated with $\bar{P}$ has a globally reachable aperiodic node, by Theorem 2.2 we have that $\mathbb{E}[x(t)]$ converges to a consensus point $c\mathbf{1}$. Moreover, the convergence rate is given by $\rho_2(\bar{P})$ and $c = v^* x(0)$, where v is the normalized dominant left eigenvector of $\bar{P}$. In principle, convergence of the expected dynamics does not, by itself, guarantee convergence of the random dynamics. However, the next result provides general and intuitive conditions for the convergence of (3.6), which are indeed based on the convergence properties of the expected dynamics.

Theorem 3.3 (Almost sure convergence to consensus) *Consider the dynamical system* (3.6) *and assume that the matrices $P(t)$ are independently and randomly sampled from an ensemble $\mathscr{P}$ of nondegenerate stochastic matrices equipped with a fixed probability distribution. Then, the following three facts are equivalent:*

(i) for every initial condition, there exists a scalar random variable x_∞ such that $x(t)$ converges almost surely to $x_\infty\mathbf{1}$;
(ii) $\rho_2(\bar{P}) < 1$;
(iii) the "expected graph" $\mathscr{G}_{\bar{P}}$ has a globally reachable node.

Proof $(i) \Rightarrow (ii)$: Being $x(t)$ a bounded sequence, if $x(t)$ converges almost surely to $x_\infty\mathbf{1}$, then also $\mathbb{E}[x(t)]$ converges to $\mathbb{E}[x_\infty]\mathbf{1}$ by the dominated convergence theorem. As $\mathbb{E}[x(t+1)] = \bar{P}\,\mathbb{E}[x(t)]$, then necessarily $\rho_2(\bar{P}) < 1$.

$(ii) \Rightarrow (iii)$: Under the assumption on the diagonal of $P(t)$, the graph $\mathscr{G}_{\bar{P}}$ has a globally reachable node if and only if $\rho_2(\bar{P}) < 1$.

$(iii) \Rightarrow (i)$: If $\mathscr{G}_{\bar{P}}$ has a globally reachable node, say $k \in V$, then there exists $m \in \mathbb{N}$ such that the kth column of $\bar{P}^m$ is positive. This implies that, for every t_0, the entry $(P(t_0 + m) \ldots P(t_0 + 1)P(t_0))_{vk}$ has a positive probability of being positive: Denote such probability by p_v. Now, recall that $P_{kk}(t) > 0$ almost surely: This implies that the kth column of matrix $P(t_0 + Nm) \ldots P(t_0 + 1)P(t_0)$ is positive with probability $\prod_{u \in V} p_u$. Consequently, there exists $\alpha > 0$ such that, with positive probability, each element of this column is larger than α. Now, let us define the sequence of times $t_h = mNh$ for $h \in \mathbb{Z}_{\geq 0}$: Reasoning as in the proof of Lemma 3.1, we can apply Lemma 2.1 to argue that

$$\max_{v \in V} x_v(t_{k+1}) - \min_{v \in V} x_v(t_{k+1}) \leq (1 - \alpha)\big(\max_{v \in V} x_v(t_k) - \min_{v \in V} x_v(t_k)\big)$$

with a positive probability which does not depend on t_k. Hence, this inequality almost surely holds for infinitely many k and then almost surely $x(t)$ converges to consensus. □

It is remarkable that Theorem 3.3 translates on the expected graph of the network the same condition for consensus that holds for time-invariant networks. Following this analogy, one would expect that the essential spectral radius of $\mathbb{E}[P(t)]$ determines the speed of convergence of the algorithm. A result in this direction can be found by a suitable mean-square analysis. Let us denote the current empirical variance as

$$x_{\text{var}}(t) := \frac{1}{N}||x(t) - x_{\text{ave}}(t)\mathbf{1}||^2 = \frac{1}{N}||\Omega x(t)||^2,$$

where $\Omega = I - \frac{1}{N}\mathbf{1}\mathbf{1}^*$, and define the mean-square rate of convergence as

$$R := \sup_{x(0)} \limsup_{t \to +\infty} \mathbb{E}[x_{\text{var}}(t)]^{1/t}. \tag{3.7}$$

Notice that

$$\mathbb{E}[x_{\text{var}}(t)] = \frac{1}{N}\mathbb{E}[x(t)^* \Omega x(t)] = \frac{1}{N} x(0)^* \Delta(t) x(0),$$

where

$$\Delta(t) := \mathbb{E}[P(0)^* P(1)^* \ldots P(t-1)\Omega P(t-1) \ldots P(1)P(0)]$$

if $t \geq 1$ and $\Delta(0) := \Omega$. Clearly,

$$\Delta(t+1) = \mathbb{E}[P(0)^* \Delta(t) P(0)].$$

This recursion shows that $\Delta(t)$ is the solution of a linear dynamical system, which can be written in the form

$$\Delta(t+1) = \mathscr{L}(\Delta(t))$$

where $\mathscr{L} : \mathbb{R}^{V\times V} \to \mathbb{R}^{V\times V}$ is given by $\mathscr{L}(M) = \mathbb{E}[P(0)^* M P(0)]$. The knowledge of the operator $\mathscr{L}$, in principle, provides all information about the mean-square analysis. For instance, we have that R is the spectral radius of the operator $\mathscr{L}$, restricted to the smallest $\mathscr{L}$-invariant subspace of $\mathbb{R}^{V\times V}$ containing Ω. This characterization, however, is not very useful, because the operator $\mathscr{L}$ is difficult to compute in the applications. The next result provides rate estimates that are easier to compute.

Proposition 3.4 (Mean-square convergence rate) *Consider* (3.6) *and the convergence rate R as in* (3.7). *Then,*

$$\rho_2(\bar{P})^2 \leq R \leq \operatorname{sr}\left(\mathbb{E}[P(t)^* \Omega P(t)]\right), \tag{3.8}$$

where we recall that $\Omega = I - N^{-1}\mathbf{1}\mathbf{1}^*$ and $\operatorname{sr}(\cdot)$ *denotes the spectral radius of a matrix.*

Proof We start from the first inequality. We define $Q(t) = P(t-1)\dots P(0)$ and notice that

$$\mathbb{E}[x^*(t)\Omega x(t)] = \mathbb{E}[||\Omega x(t)||^2] = \mathbb{E}[||\Omega Q(t)x(0)||^2].$$

Now using Jensen's inequality, we have that

$$\mathbb{E}[||\Omega Q(t)x(0)||^2] \geq ||\mathbb{E}[\Omega Q(t)x(0)]||^2 = ||\Omega \bar{P}^t x(0)||^2,$$

which proves the inequality.

In order to prove the second inequality, let $y \in \mathbb{R}^V$ and note

$$\begin{aligned} y^*\mathbb{E}[P(0)^*\Omega P(0)]y &= \mathbb{E}[y^*\Omega P(0)^*\Omega P(0)\Omega y] \\ &= y^*\Omega\mathbb{E}[P(0)^*\Omega P(0)]\Omega y \\ &\leq ||\mathbb{E}[P(0)^*\Omega P(0)]||y^*\Omega y, \end{aligned}$$

by the symmetry of the matrix. We deduce that $\mathscr{L}(\Omega) \leq ||\mathscr{L}(\Omega)||\Omega$. This fact, together with the remark that if $M_1 \leq M_2$, then $\mathscr{L}(M_1) \leq \mathscr{L}(M_2)$, implies that

$$\mathscr{L}^t(\Omega) = \mathscr{L}^{t-1}(\mathscr{L}(\Omega)) \leq \mathscr{L}^{t-1}(||\mathscr{L}(\Omega)||\Omega) = ||\mathscr{L}(\Omega)||\mathscr{L}^{t-1}(\Omega).$$

By iterating this reasoning, we get $\mathscr{L}^t(\Omega) \leq ||\mathscr{L}(\Omega)||^t\Omega$, which gives the thesis. □

Note that if all matrices $P(t)$ are symmetric, then $\mathbb{E}[P(t)^*\Omega P(t)] = \mathbb{E}[P^2(t)] - \frac{1}{N}\mathbf{1}\mathbf{1}^*$. In the special case of the USG algorithm, we further have $\mathbb{E}[P^2(t)] - \frac{1}{N}\mathbf{1}\mathbf{1}^* = \mathbb{E}[P(t)] - \frac{1}{N}\mathbf{1}\mathbf{1}^*$ and we can thus argue that $\rho_2(\mathbb{E}[P(t)])^2 \leq R \leq \rho_2(\mathbb{E}[P(t)])$.

We now introduce two examples of randomized averaging algorithms, which can be studied by the above results. The first example generalizes the USG and features the activation of *one pair* of connected nodes per time step: The two nodes communicate with each other and both update their states.

Example 3.3 (*symmetric gossip algorithm (SG)*) Let a weighted graph $G=(I, E, W)$ and $q \in (0, 1)$ be given, such that W is *symmetric* and $\mathbf{1}^* W \mathbf{1} = 1$. For every $t \geq 0$, one edge $(v, w) \in E$ is sampled from a distribution such that the probability of selecting (v, w) is W_{vw}. Then,

$$\begin{aligned} x_v(t+1) &= (1-q)\, x_v(t) + q\, x_w(t) \\ x_w(t+1) &= (1-q)\, x_w(t) + q\, x_v(t) \\ x_u(t+1) &= x_u(t) \quad \text{for } u \neq v, w. \end{aligned}$$

Both W and q can be considered in principle as design parameters, with respect to which one can optimize the performance.

In order to analyze the SG algorithm, for every (v, w) we let

$$P^{(v,w)} := I - q(e_v - e_w)(e_v - e_w)^* = I - q(e_v e_v^* - e_w e_v^* - e_v e_w^* + e_w e_w^*),$$

where e_u is the uth vector of the canonical basis of $\mathbb{R}^V$. Note that trivially $W = \sum_{(v,w)} W_{vw} e_v e_w^*$. Then, the distribution of $P(t)$ is concentrated on these matrices and $\mathbb{P}[P(t) = P^{(v,w)}] = W_{vw}$. We have that (using the notation for the Laplacian of a matrix)

$$\begin{aligned} \mathbb{E}[P(t)] &= \sum_{(v,w)} W_{vw} P^{(v,w)} \\ &= I - q \sum_{(v,w)} W_{vw}(e_v - e_w)(e_v - e_w)^* \\ &= I - 2qL(W). \end{aligned}$$

Note that if the graph associated with W is strongly connected, then the average graph is automatically strongly connected. Since in the SG all the diagonal elements of $P(t)$ are nonzero with probability 1 and all the $P(t)$s are symmetric, we can apply Theorem 3.3 and conclude that this algorithm yields average consensus almost surely. Moreover, noting that $\|e_v - e_w\|_2^2 = 2$ and

$$\begin{aligned} (P^{(v,w)})^2 &= I - 2q(e_v - e_w)(e_v - e_w)^* + q^2(e_v - e_w)(e_v - e_w)^*(e_v - e_w)(e_v - e_w)^* \\ &= I - 2q(1-q)(e_v - e_w)(e_v - e_w)^*, \end{aligned}$$

we argue that

$$\mathbb{E}[P(t)^* \Omega P(t)] = \mathbb{E}[P(t)^2] - \frac{1}{N}\mathbf{1}\mathbf{1}^* = \Omega - 4q(1-q)L(W).$$

Then, by applying Proposition 3.4, we can estimate the convergence rate as

$$\rho_2(I - 2qL(W))^2 \le R \le \mathrm{sr}(\Omega - 4q(1-q)L(W))$$

and, provided we denote by λ the smallest nonzero eigenvalue of $L(W)$, as

$$1 - 4q\lambda \le R \le 1 - 4q(1-q)\lambda.$$

The next example features the activation of *one node* per time step. The activated node communicates its current state to all its neighbors, which in turn update their states. We note that the algorithm is inherently asymmetric: As a consequence, the average of the initial states is not preserved.

Example 3.4 (*Broadcast gossip algorithm (BG)*) Let there be $q \in (0, 1)$ and a directed graph $G = (V, E)$ whose adjacency matrix is denoted by $A \in \{0, 1\}^{V \times V}$. For every $t \ge 0$, one node w is sampled from a uniform distribution over V. Then,

$$\begin{aligned} x_v(t+1) &= (1-q)\, x_v(t) + q\, x_w(t) && \text{if } A_{vw} > 0 \\ x_v(t+1) &= x_v(t) && \text{otherwise.} \end{aligned}$$

In other words, one randomly selected node broadcasts its value to all its neighbors, which update their values accordingly.

For the analysis of this algorithm, we define

$$P^{(w)} = I - q \sum_{v: A_{vw} > 0} (e_v e_v^* - e_v e_w^*)$$

and note that $\mathbb{P}[P(t) = P^{(w)}] = \frac{1}{N}$. Then,

$$\mathbb{E}[P(t)] = I - \frac{q}{N} L.$$

If the graph G is strongly connected, then the algorithm converges to consensus almost surely. Before we further investigate the properties of the BG algorithm, we assume that the graph is topologically balanced, i.e., $A\mathbf{1} = A^*\mathbf{1}$. This property in particular implies that $\mathbf{1}^* L = \mathbf{1}^*$ and then $\mathbb{E}[x_\infty] = x_{\mathrm{ave}}(0)$. Moreover, the reader may compute that

$$\mathbb{E}[P(t)^* P(t)] = I - \frac{q(1-q)}{N}(L + L^*) \tag{3.9a}$$

$$\mathbb{E}[P(t)^* \mathbf{1}\mathbf{1}^* P(t)] = \mathbf{1}\mathbf{1}^* + \frac{q^2}{N} L L^*. \tag{3.9b}$$

As a consequence, the convergence rate can be estimated using Proposition 3.4 as

$$\rho_2\left(I - \frac{q(1-q)}{N}(L + L^*)\right)^2 \le R \le \mathrm{sr}\left(\Omega - \frac{q(1-q)}{N}(L + L^*) - \frac{q^2}{N^2} L L^*\right)$$

If we denote by 2λ the smallest nonzero eigenvalue of $L + L^*$, and we remark that LL^* is positive semidefinite, the above bounds can be simplified to

$$1 - 4\frac{q(1-q)}{N}\lambda \le R \le 1 - 2\frac{q(1-q)}{N}\lambda.$$

We have seen that, provided the graph is balanced, the BG algorithm yields $\mathbb{E}[x_\infty] = x_{\text{ave}}(0)$. Considering that x_∞ is a random variable, its spreading around the mean value needs to be evaluated. To this aim, we introduce the mean-square error $\mathbb{E}\left[(x_\infty - x_{\text{ave}}(0))^2\right]$ and below we provide a technical tool to estimate it. In order to state the result, it is again convenient to denote the empirical variance as

$$x_{\text{var}}(t) := \frac{1}{N}\sum_{i=1}^{N}\big(x_i(t) - x_{\text{ave}}(t)\big) = \frac{1}{N}x^*\Omega x(t).$$

Theorem 3.4 (Accuracy condition) *Consider dynamics* (3.6) *and assume that* $\mathbf{1}^*\bar{P} = \mathbf{1}^*$ *and that there exists* $\gamma > 0$ *such that*[1]

$$\mathbb{E}[P^*\mathbf{1}\mathbf{1}^*P] - \mathbf{1}\mathbf{1}^* \le \gamma\big(I - \mathbb{E}[P^*P]\big). \tag{3.10}$$

Then,

$$\mathbb{E}\left[(x_{\text{ave}}(t) - x_{\text{ave}}(0))^2\right] \le \frac{\gamma}{N+\gamma}\mathbb{E}\left[x_{\text{var}}(0) - x_{\text{var}}(t)\right]. \tag{3.11}$$

If additionally $\mathcal{G}_{\bar{P}}$ *has a globally reachable node, then*

$$\mathbb{E}\left[(x_\infty - x_{\text{ave}}(0))^2\right] \le \frac{\gamma}{N+\gamma}x_{\text{var}}(0). \tag{3.12}$$

Proof In the proof, we shall use the notation $x_{\text{ave}} = N^{-1}\mathbf{1}^*x$ and $x_{\text{var}} = N^{-1}x^*\Omega x$ to denote the empirical average and variance of a generic vector $x \in \mathbb{R}^V$. We let

$$\begin{aligned} C(x) &:= N(\gamma+N)x_{\text{ave}}^2 + N\gamma\, x_{\text{var}} \\ &= \frac{N(\gamma+N)}{N^2}x^*\mathbf{1}\mathbf{1}^*x + \frac{N\gamma}{N}x^*\Big(I - \frac{1}{N}\mathbf{1}\mathbf{1}^*\Big)x \\ &= x^*\Big(\mathbf{1}\mathbf{1}^* + \gamma\, I\Big)x. \end{aligned}$$

Then, for a generic stochastic matrix P, we have that

$$C(Px) - C(x) = x^*\Big(P^*\mathbf{1}\mathbf{1}^*P + \gamma P^*P - \mathbf{1}\mathbf{1}^* - \gamma I\Big)x.$$

---

[1] In this result, inequalities between matrices like $A \le B$ have to be intended as $A - B$ being negative semidefinite.

Consequently, condition (3.10) implies that for dynamics (3.6)

$$\mathbb{E}[C(x(t+1)) - C(x(t))|x(t)] = x(t)^* \Big(\mathbb{E}[P^*\mathbf{1}\mathbf{1}^*P + \gamma\, P^*P] - \mathbf{1}\mathbf{1}^* - \gamma\, I\Big)x(t) \leq 0$$

and $\mathbb{E}[C(x(t))] \leq C(x(0))$ for all $t \in \mathbb{N}$. This inequality can be rewritten as

$$\mathbb{E}\left[x_{\text{ave}}(t)^2 - x_{\text{ave}}(0)^2\right] \leq \frac{\gamma}{N+\gamma}\mathbb{E}\left[x_{\text{var}}(0) - x_{\text{var}}(t)\right].$$

This inequality implies (3.11) if $x_{\text{ave}}(0) = 0$: The general case follows by applying this special case to the translated dynamics $x - x_{\text{ave}}(0)\mathbf{1}$. Finally, inequality (3.12) is an immediate corollary of convergence. □

In the case of the broadcast gossip algorithm, (3.9) implies that (3.10) reads

$$\frac{q^2}{N}LL^* \leq \gamma\,\frac{q(1-q)}{N}(L+L^*),$$

which holds true for $\gamma = d_{\max}\frac{q}{1-q}$ because for balanced graphs $LL^* \leq d_{\max}(L+L^*)$. Consequently,

$$\mathbb{E}\left[(x_\infty - x_{\text{ave}}(0))^2\right] \leq \frac{q}{1-q}\frac{d_{\max}}{N}x_{\text{var}}(0).$$

Remarkably, as long as $d_{\max} = o(N)$, this upper bound goes to zero as N goes to infinity, that is, *the error committed by the algorithm in approximating the average becomes negligible on large networks.*

Exercises

Exercise 3.1 (*Strong positivity* [35]) Consider the sequence of matrices

$$P(t) = \begin{pmatrix} 1-\alpha_t & \alpha_t \\ \alpha_t & 1-\alpha_t \end{pmatrix}$$

where $\alpha_t \in [0, 1]$ is a given sequence. Observe that all $P(t)$ are aperiodic and irreducible.

(i) Prove that, if $\alpha_t \geq \alpha > 0$ for all t, then the sequence $P(t)$ leads to a consensus.
(ii) For sequences $\alpha_t \to 0$ when $t \to +\infty$, find sufficient conditions on the speed of convergence which guarantee that $P(t)$ leads to a consensus.
(iii) Find an explicit example of a sequence $\alpha_t \to 0$ for $t \to +\infty$, for which $P(t)$ does not lead to a consensus.

Exercise 3.2 (*Time-varying consensus on the line graph*) Let $V = \{0, \ldots, N-1\}$ and consider the directed line graph $\mathbf{L} = (V, E)$. Let the vector $e_u \in \mathbb{R}^V$ be such that the vth component of e_u is equal to 1 if $v = u$ and to 0 otherwise, and define the matrix $P^{(u,v)} = I - \frac{1}{2}(e_u e_u^* - e_u e_v^*)$. Consider a time-dependent consensus algorithm (3.1) with

$$P(t) = P^{(k,k+1)} \quad \text{where } k = t \pmod{N}.$$

(i) Verify that the dynamics satisfies the assumptions of Theorem 3.1, and find the minimal T for the connectivity assumption.
(ii) Verify that the dynamics satisfies the assumptions of Lemma 3.1, finding the suitable value of B. Compare B with the value of T found in (i).

Exercise 3.3 (*Uniform connectivity* [35]) Consider (3.1) with $x(0) = (0, 1, 1)^*$ and the sequence $\{P(t)\}_t$ defined as follows. Let

$$P_1 = \begin{bmatrix} 1 & 0 & 0 \\ 1/2 & 1/2 & 0 \\ 0 & 0 & 1 \end{bmatrix} \qquad P_2 = \begin{bmatrix} 1/2 & 1/2 & 0 \\ 1/2 & 1/2 & 0 \\ 0 & 0 & 1 \end{bmatrix}$$

$$P_3 = \begin{bmatrix} 1 & 0 & 0 \\ 0 & 1/2 & 1/2 \\ 0 & 0 & 1 \end{bmatrix} \qquad P_4 = \begin{bmatrix} 1 & 0 & 0 \\ 0 & 1/2 & 1/2 \\ 0 & 1/2 & 1/2 \end{bmatrix}$$

and $Q_s = \underbrace{P_1, \dots, P_1}_{2s}, P_2, \underbrace{P_3, \dots, P_3}_{2s+1}, P_4$. Assume the sequence $P(t)$ is the concatenation of $Q_0, Q_1, Q_2, \dots$. Then, show that $x(t)$ does not converge to a consensus.

Exercise 3.4 (*Krause's convergence time*) Consider Krause's dynamics (3.5).

(i) Show that the order between opinions is preserved, i.e., for all $t \geq 0$, if $x_v(t) \leq x_w(t)$, then $x_v(t+1) \leq x_w(t+1)$. This implies that we can assume (without loss of generality) that the agents are sorted, i.e., if $v < w$, then $x_v < x_w$.
(ii) Show that if at some time t the distance between two consecutive agent opinions $x_v(t)$ and $x_{v+1}(t)$ is larger than ε, then it remains so for all time $s > t$.
(iii) Assume from now on that $\rho = 1$. Show that there exist $T \in \mathbb{N}$ and $\tilde{x} \in \mathbb{R}^V$ such that $x(t) = \tilde{x}$ for all $t \geq T$.
(iv) Show that for all $v, w \in V$, either $|\tilde{x}_v - \tilde{x}_w| > \varepsilon$ or $\tilde{x}_v = \tilde{x}_w$.
(v) Estimate the worst-case converge time $\bar{T} = \sup_{x(0)} \inf\{t : x(t) = \tilde{x}\}$ ([5, Sect. 4.6.1]).

Exercise 3.5 (*Unbounded confidence*) Consider the following generalized Krause's model. Fix a continuous function $\xi : [0, +\infty) \to [0, +\infty)$ such that $\xi(x) > 0$ for all $x \geq 0$, and define

$$x_v(t+1) = x_v(t) + \frac{\rho}{\sum\limits_{w \in V} \xi(|x_w(t) - x_v(t)|)} \sum_{w \in V} \xi(|x_w(t) - x_v(t)|)(x_w(t) - x_v(t)), \tag{3.13}$$

where $\rho \in (0, 1)$. Fix some initial condition $x(0)$ and let $P(t)$ be the sequence of matrix such that $x(t+1) = P(t)x(t)$.

(i) Prove that the sequence $P(t)$ is nondegenerate and conclude, using Corollary 3.2, that it leads to a consensus.
(ii) Assume that $\xi(x) = e^{-x^2}$ and let $d(t) = \max\{x_u(t)\} - \min\{x_u(t)\}$. Using Lemma 2.1, prove that

$$d(t+1) \leq \left(1 - e^{-d(t)^2}\right) d(t)$$

(iii) Assuming that $x(0)_v \in [-1, 1]$ for all $v \in V$, find, for fixed $\varepsilon > 0$, an estimate of the convergence time

$$t_\varepsilon := \inf\{t \mid d(t) \leq \varepsilon\}$$

Compare this estimate with explicit simulations of (3.13) for $N = 10$.

Exercise 3.6 (*Broadcast on a star*) Let $S_N = (V, E)$ be a star with N leaves. Consider the following randomized consensus algorithm. For all positive integers t, sample one node w from a uniform distribution over V, and update the states as follows:

$$\begin{aligned} x_v(t+1) &= (1-q)x_v(t) + qx_w(t) \qquad \text{if } (v, w) \in E \\ x_v(t+1) &= x_v(t) \qquad \text{if } (v, w) \notin E. \end{aligned}$$

The update parameter satisfies $q \in (0, 1)$.

(i) Write down the update rule for the proposed algorithm in matrix form.
(ii) Compute the expected update matrix $\bar{P} = \mathbb{E}[P(t)]$.
(iii) Let $x_{\text{ave}}(t) = \frac{1}{N}\sum_{v \in V} x_v(t)$. Verify that, although in general $x_{\text{ave}}(t+1) \neq x_{\text{ave}}(t)$, nevertheless $\mathbb{E}[x_{\text{ave}}(t+1)] = \mathbb{E}[x_{\text{ave}}(t)]$.
(iv) Show that the algorithm ensures almost sure convergence of the states.
(v) Compute the second largest eigenvalue of $\bar{P}$. To this goal, you may use Exercise 2.18.
(vi) Estimate the convergence rate R of the algorithm as a function of q and N, and conclude that $\lim_{N\to\infty} R = 1$ irrespective of q.

Exercise 3.7 (*Triplet-gossip*) Consider the following random dynamics on a complete graph $G = (V, E)$. At every discrete-time step t, three agents u, v, w are uniformly and independently sampled from V, and they update their internal state as follows:

$$x_u(t+1) = x_v(t+1) = x_w(t+1) = \frac{x_u(t) + x_v(t) + x_w(t)}{3}$$

Let $P(t)$ be the corresponding matrix acting on the full vector $x(t)$.

(i) Compute $\overline{P} = \mathbb{E}[P(t)]$, its eigenvalues, and its spectral gap.
(ii) Give an estimation of $\mathbb{E}||\Omega x(t)||^2$ analogous to what is done for the pairwise gossip algorithm in Example 3.3.

Exercise 3.8 (*Asynchronous asymmetric gossip algorithm (AAGA)*) This exercise gives an example of randomized algorithm, in which one *directed* edge is activated at each time step, resulting in an asymmetric update rule.

Let a weighted graph $G = (V, E, W)$ and $q \in (0, 1)$ be given, such that $W\mathbf{1} = W^*\mathbf{1}$ and $\mathbf{1}^* W \mathbf{1} = 1$. For every $t \geq 0$, one edge (v, w) is sampled from a distribution such that the probability of selecting (v, w) is W_{vw}. Then, we define

$$x_v(t+1) = (1-q)\, x_v(t) + q\, x_w(t) \tag{3.14}$$

and $x_u(t+1) = x_u(t)$ for $u \neq v$.

(i) Verify that (as proved in [18, Sect. 4])

$$\begin{aligned} \mathbb{E}[P(t)] &= I - qL(W) \\ \mathbb{E}[P(t)^* P(t)] &= I - q(1-q)L(W + W^*) \\ \mathbb{E}[P(t)^* \mathbf{1}\mathbf{1}^* P(t)] &= \mathbf{1}\mathbf{1} + q^2 L(W + W^*). \end{aligned}$$

(ii) Assume from now on that the graph G is strongly connected. Prove that system (3.14) almost surely converges to a limit value x_∞ such that $\mathbb{E}[x_\infty] = x_{\text{ave}}(0)$ and

$$\mathbb{E}\left[(x_\infty - x_{\text{ave}}(0))^2\right] \leq \frac{q}{1-q}\frac{1}{N} x_{\text{var}}(0).$$

(iii) Assume moreover that W is symmetric and denote by λ the spectral gap of $L(W)$. Show that the convergence rate is bounded by

$$1 - 2q\lambda \leq R \leq 1 - 2q\left((1-q) + \frac{q}{N}\right)\lambda,$$

provided N is large enough.

Bibliographical Notes

Deterministic networks. Our choice of results on deterministic time-dependent consensus mostly consist of necessary and sufficient conditions for convergence. While the convergence analysis of time-invariant averaging dates back at least to De Groot [13] in 1974, sufficient conditions for the time-varying case were given by [45] in 1984 and later by [7, 8, 28, 31]. The results on convergence to consensus that we present in Sect. 3.1 appeared, in quite a different formulation, in [35]: The counterexample in Exercise 3.3 is in the original paper. Our version of the results is based on the analysis in [24, 27].

In consensus-seeking systems, results which do make a "global" assumption of connectivity can ensure convergence but possibly not consensus. Such results are motivated by the difficulty to satisfy, in some applications, connectivity conditions over time. In this spirit, we have presented Theorem 3.2, which is more general than early results as those in [31]. The result can be found in [25, Theorem 2]: We present it here with a proof which is adapted to the rest of our arguments. The works [26, 27] developed the notion of *cut-balance*, which has been used and extended in several subsequent works, including [11, 33, 42, 44]. However, it is clear that in certain applications, e.g., vehicle rendezvous, mere convergence is not satisfactory. For this reason, there has been much work devoted to variations of the consensus algorithm, which inherently guarantee connectivity. A discussion about this *connectivity maintenance* issue may be found, for instance, in [5, Chap. 4]. In the opposite direction, Krause's model is a simple but very interesting example of a consensus-seeking dynamics without a global connectivity assumption. The dynamics was originally proposed in [23, 29], as a model for opinion dynamics with bounded confidence [32]. Krause's dynamics have been the topic of several works [2], which have also considered variations of the dynamics that feature continuous-time evolution [3, 9], multi-dimensional opinions [16, 38], heterogenous thresholds [34], and continua of agents [3, 6].

Instead, we did not investigate much two important issues that have been discussed in detail in Chap. 2 for time-invariant networks: speed of convergence and limit state. Studying the speed of convergence of time-dependent consensus algorithms is indeed quite delicate. First, it is essential to assume connectedness on bounded interval; otherwise, the algorithm can be arbitrarily slowed down by introducing arbitrary sequences of disconnected graphs. Moreover, even if connectivity on bounded intervals is assumed, the convergence time can be large (as in Exercise 3.2), even exponentially large in the interval size and in N: We refer to [37] for a detailed discussion. Results have been proved for specific dynamics, such as Krause's, see Exercise 3.4 and [15]. Stronger results can be found by assuming the matrices $P(t)$ to be nondegenerate and their associated graphs connected at each time step: Recent results in this framework can be found in [36]. Also the issue of determining the consensus (or convergence) point has no simple answer in the literature. In principle, such an analysis reduces to studying the *absolute probability vectors* as defined in [39]. Explicit results, however, are only available in special cases: For instance, see [2] for a partial description of the limit states of Krause's dynamics and [46] for some recent developments on this matter.

Randomized networks. Theorem 3.3 was originally proven in [19, 40]. The proof presented here is new, although inspired by [40], and has been written to seamlessly take advantage of our treatment of the deterministic case. We acknowledge that Theorem 3.3 is not the most general convergence result for randomized consensus dynamics, because it requires positivity of the diagonal and statistical independence of the update matrices. This independence assumption can be significantly relaxed, at the price of using more subtle probabilistic tools. In [41], the condition $\rho_2(\mathbb{E}[P(t)]) < 1$ is proven to be necessary and sufficient for consensus, under the more general assumption that the sequence $P(t)$ is generated by an ergodic stationary process (and has positive diagonals). In [30], it is proved that in fact any adapted stochastic process is suitable, provided certain assumptions of uniform connectivity hold. Also the assumption of positivity of the diagonal can be relaxed. A necessary and sufficient condition for convergence, which does not require to assume that the diagonal is positive, can be found in [19, Theorem 3.1] and is due to Cogburn [10]. An intermediate useful condition is *strong aperiodicity* as defined in [42]. Very general conditions for adapted processes of "balanced" stochastic matrices are provided in [43].

For randomized dynamics, we have presented some estimates on the speed of convergence, by using the mean-square analysis developed in [19]. Another important topic is studying the random variable x_∞ and its distance from $x_{\text{ave}}(0)$. Obtaining a complete characterization of the distribution of x_∞ seems to remain an open problem, but a few practical results are available to estimate its variance. In principle, the variance of the consensus value can be exactly computed by the formula in [41, Eq. (7)], which involves the dominant eigenvectors of the first two moments of the update matrix: However, this characterization can be inconvenient in the applications and does not provide clear insights on the scaling for large networks. More recently, an effective estimate has been derived in [22], providing conditions for the variance to go to zero as N goes to infinity. This result, which we presented in Theorem 3.4, covers a wide class of randomized algorithms that involve asymmetric communication or packet losses [20, 21].

There is a large variety of examples in randomized consensus dynamics, and our selection has focused on two models which have possibly been the most popular in recent literature [14]. The symmetric gossip algorithm in Example 3.3 was popularized in the systems and control community by the influential work [4]. The broadcast gossip algorithm in Example 3.4 has attracted a significant attention, because it involves broadcast communication and thus applies very naturally to wireless networks [1, 17, 19]: Formulas (3.9) can be found in [1, Lemma 4]. As we have shown, the consensus value of BG does not coincide with the average of the initial conditions, but this bias becomes negligible for large networks [22]. Another intuitive algorithm with the same property is the asymmetric gossip algorithm [18], which we introduce in Exercise 3.8. Finally, we would like to mention that there exist dynamics that combine gossiping and bounded confidence, proposed as opinion dynamics models [12, 47].

References

1. Aysal, T.C., Yildiz, M.E., Sarwate, A.D., Scaglione, A.: Broadcast gossip algorithms for consensus. IEEE Trans. Signal Process. **57**(7), 2748–2761 (2009)
2. Blondel, V.D., Hendrickx, J.M., Tsitsiklis, J.N.: On Krause's multi-agent consensus model with state-dependent connectivity. IEEE Trans. Autom. Control **54**(11), 2586–2597 (2009)
3. Blondel, V.D., Hendrickx, J.M., Tsitsiklis, J.N.: Continuous-time average-preserving opinion dynamics with opinion-dependent communications. SIAM J. Control Optim. **48**(8), 5214–5240 (2010)
4. Boyd, S., Ghosh, A., Prabhakar, B., Shah, D.: Randomized gossip algorithms. IEEE Trans. Inf. Theory **52**(6), 2508–2530 (2006)

5. Bullo, F., Cortés, J., Martínez, S.: Distributed Control of Robotic Networks. Applied Mathematics Series. Princeton University Press (2009)
6. Canuto, C., Fagnani, F., Tilli, P.: An Eulerian approach to the analysis of Krause's consensus models. SIAM J. Control Optim. **50**(1), 243–265 (2012)
7. Cao, M., Morse, A.S., Anderson, B.D.O.: Agreeing asynchronously. IEEE Trans. Autom. Control **53**(8), 1826–1838 (2008)
8. Cao, M., Morse, A.S., Anderson, B.D.O.: Reaching a consensus in a dynamically changing environment - convergence rates, measurement delays and asynchronous events. SIAM J. Control Optim. **47**(2), 601–623 (2008)
9. Ceragioli, F., Frasca, P.: Continuous and discontinuous opinion dynamics with bounded confidence. Nonlinear Anal. Real World Appl. **13**(3), 1239–1251 (2012)
10. Cogburn, R.: On products of random stochastic matrices. In: Random Matrices and their Applications, vol. 50. Contemporary Mathematics, pp. 199–213. American Mathematical Society (1986)
11. De Persis, C., Frasca, P., Hendrickx, J.M.: Self-triggered rendezvous of gossiping second-order agents. In: IEEE Conference on Decision and Control, Florence, Italy, pp. 7403–7408 (2013)
12. Deffuant, G., Neau, D., Amblard, F., Weisbuch, G.: Mixing beliefs among interacting agents. Adv. Complex Syst. **3**(1–4), 87–98 (2000)
13. DeGroot, M.H.: Reaching a consensus. J. Am. Stat. Assoc. **69**(345), 118–121 (1974)
14. Dimakis, A.G., Kar, S., Moura, J.M.F., Rabbat, M.G., Scaglione, A.: Gossip algorithms for distributed signal processing. Proc. IEEE **98**(11), 1847–1864 (2010)
15. Etesami, S.R.: Potential-based analysis of social, communication, and distributed networks. Ph.D. thesis, University of Illinois at Urbana-Champaign (2015)
16. Etesami, S.R., Başar, T.: Game-theoretic analysis of the Hegselmann-Krause model for opinion dynamics in finite dimensions. IEEE Trans. Autom. Control **60**(7), 1886–1897 (2015)
17. Fagnani, F., Frasca, P.: Broadcast gossip averaging: interference and unbiasedness in large Abelian Cayley networks. IEEE J. Sel. Top. Signal Process. **5**(4), 866–875 (2011)
18. Fagnani, F., Zampieri, S.: Asymmetric randomized gossip algorithms for consensus. In: IFAC World Congress, Seoul, Korea, pp. 9052–9056, July 2008
19. Fagnani, F., Zampieri, S.: Randomized consensus algorithms over large scale networks. IEEE J. Sel. Areas Commun. **26**(4), 634–649 (2008)
20. Fagnani, F., Zampieri, S.: Average consensus with packet drop communication. SIAM J. Control Optim. **48**(1), 102–133 (2009)
21. Frasca, P., Hendrickx, J.M.: Large network consensus is robust to packet losses and interferences. In: European Control Conference, Zürich, Switzerland, pp. 1782–1787, July 2013
22. Frasca, P., Hendrickx, J.M.: On the mean square error of randomized averaging algorithms. Automatica **49**(8), 2496–2501 (2013)
23. Hegselmann, R., Krause, U.: Opinion dynamics and bounded confidence: models, analysis and simulation. J. Artif. Soc. Soc. Simul. **5**(3), 1–33 (2002)
24. Hendrickx, J.M.: Graphs and Networks for the Analysis of Autonomous Agent Systems. Ph.D. thesis, Université catholique de Louvain (2008)
25. Hendrickx, J.M., Tsitsiklis, J.N.: Convergence of type-symmetric and cut-balanced consensus seeking systems (extended version). http://arxiv.org/abs/1102.2361 (2011)
26. Hendrickx, J.M., Tsitsiklis, J.N.: A new condition for convergence in continuous-time consensus seeking systems. In: IEEE Conference on Decision and Control and European Control Conference, Orlando, FL, USA, pp. 5070–5075 (2011)
27. Hendrickx, J.M., Tsitsiklis, J.N.: Convergence of type-symmetric and cut-balanced consensus seeking systems. IEEE Trans. Autom. Control **58**(1), 214–218 (2013)
28. Jadbabaie, A., Lin, J., Morse, A.S.: Coordination of groups of mobile autonomous agents using nearest neighbor rules. IEEE Trans. Autom. Control **48**(6), 988–1001 (2003)
29. Krause, U.: A discrete nonlinear and non-autonomous model of consensus formation. In: Elaydi, S., Ladas, G., Popenda, J., Rakowski, J. (eds.) Communications in Difference Equations. Gordon & Breach, pp. 227–236 (2000)

30. Liu, B., Lu, W., Chen, T.: Consensus in networks of multiagents with switching topologies modeled as adapted stochastic processes. SIAM J. Control Optim. **49**(1), 227–253 (2011)
31. Lorenz, J.: A stabilization theorem for continuous opinion dynamics. Phys. A **355**(1), 217–223 (2005)
32. Lorenz, J.: Continuous opinion dynamics under bounded confidence: A survey. Int. J. Mod. Phys. C **18**(12), 1819–1838 (2007)
33. Martin, S., Hendrickx, J.M.: Continuous-time consensus under non-instantaneous reciprocity. IEEE Trans. Autom. Control **61**(9), 2484–2495 (2016)
34. Mirtabatabaei, A., Bullo, F.: Opinion dynamics in heterogeneous networks: convergence conjectures and theorems. SIAM J Control Optim. **50**(5), 2763–2785 (2012)
35. Moreau, L.: Stability of multiagent systems with time-dependent communication links. IEEE Trans. Autom. Control **50**(2), 169–182 (2005)
36. Nedich, A., Liu, J.: Lyapunov approach to consensus problems (2014)
37. Olshevsky, A., Tsitsiklis, J.N.: Convergence speed in distributed consensus and averaging. SIAM J. Control Optim. **48**(1), 33–55 (2009)
38. Roozbehani, M., Megretski, A., Frazzoli, E.: Lyapunov analysis of quadratically symmetric neighborhood consensus algorithms. In: IEEE Conference on Decision and Control, Cancún, México, pp. 2252–2257 (2008)
39. Seneta, E.: Non-negative Matrices and Markov Chains. Springer Series in Statistics. Springer (2006)
40. Tahbaz-Salehi, A., Jadbabaie, A.: A necessary and sufficient condition for consensus over random networks. IEEE Trans. Autom. Control **53**(3), 791–795 (2008)
41. Tahbaz-Salehi, A., Jadbabaie, A.: Consensus over ergodic stationary graph processes. IEEE Trans. Autom. Control **55**(1), 225–230 (2010)
42. Touri, B.: Product of Random Stochastic Matrices and Distributed Averaging. Springer Theses. Springer, March 2012
43. Touri, B., Langbort, C.: On endogenous random consensus and averaging dynamics. IEEE Trans. Autom. Control **1**(3), 241–248 (2014)
44. Touri, B., Nedic, A.: Product of random stochastic matrices. IEEE Trans. Autom. Control **59**(2), 437–448 (2014)
45. Tsitsiklis, J.N.: Problems in Decentralized Decision Making and Computation. Ph.D. thesis, Massachusetts Institute of Technology, November 1984
46. Wang, C., Li, Q., Weinan, E., Chazelle, B.: Noisy Hegselmann-Krause systems: phase transition and the 2r-conjecture. J. Stat. Phys. **166**(5), 1209–1225 (2017)
47. Zhang, J., Hong, Y.: Opinion evolution analysis for short-range and long-range Deffuant-Weisbuch models. Phys. A **392**(21), 5289–5297 (2013)

Chapter 4
Performance and Robustness of Averaging Algorithms

Abstract This chapter has the goal of introducing more instruments for the study of consensus algorithms. We will define several performance metrics: Each proposed metric highlights a specific aspect of the algorithm, possibly in relation with a field of application. Namely, we shall consider the speed of convergence in Sect. 4.1, a quadratic control cost in Sect. 4.2, the robustness to noise in Sect. 4.3, and the estimation error in a distributed inference problem in Sect. 4.5. The metrics that we describe share the following feature: under suitable assumptions of symmetry of the update matrix, they can be evaluated as functions of the eigenvalues of the update matrix.

4.1 A Deeper Analysis of the Convergence to Consensus

We consider the usual time-invariant consensus dynamics

$$x(t+1) = Px(t), \tag{4.1}$$

where the matrix P is adapted to a strongly connected aperiodic graph $G = (V, E)$ of order N. For simplicity, we assume that the matrix P is *symmetric*, although this assumption can be relaxed to some extent. The eigenvalues of P are denoted as μ_i for $i \in \{1, \dots, N\}$ and $\mu_1 = 1$. We recall that the second eigenvalue is defined as $\rho_2 = \max\{|\mu_i|,\ i = 2, \dots, N\}$. When convenient, we will also make suitable assumptions on the statistics of the initial condition.

In Chap. 2, the speed of convergence to the consensus value of dynamics (4.1) has been estimated in terms of the second eigenvalue ρ_2 of the matrix P. From Proposition 2.2 and Corollary 2.3, we can recall that the second eigenvalue determines the convergence rate according to the estimate

$$N^{-1}||P^t x(0) - N^{-1}\mathbf{1}\mathbf{1}^* x(0)||^2 \leq \rho_2^{2t} N^{-1}||x(0)||. \tag{4.2}$$

© Springer International Publishing AG 2018

F. Fagnani and P. Frasca, *Introduction to Averaging Dynamics over Networks*, Lecture Notes in Control and Information Sciences 472,
https://doi.org/10.1007/978-3-319-68022-4_4

Notice that we have multiplied both sides of the inequality by N^{-1}. Indeed, in the large-scale limit $N \to +\infty$, it makes sense to consider the normalized version of the squared norm $N^{-1}||\cdot||^2$, because $|x(0)_v| \leq \epsilon$ for every $v \in V$ yields $N^{-1}||x(0)||^2 \leq \epsilon^2$.

There are applications, however, where estimate (4.2) turns out to be too loose or simply not adequate to the specific context. We consider the simplest such case, where the initial conditions $x(0)_v$ are assumed to be realizations of independent random variables with mean m and variance σ^2. From (4.2), by taking the mean value we obtain:

$$N^{-1}\mathbb{E}||P^t x(0) - N^{-1}\mathbf{1}\mathbf{1}^* x(0)||^2 \leq \sigma^2 \rho_2^{2t}. \tag{4.3}$$

Actually, in this case, it is possible to work out an exact characterization of the mean distance:

$$\begin{aligned}
\frac{1}{N}\mathbb{E}||P^t x(0) - N^{-1}\mathbf{1}\mathbf{1}^* x(0)||^2 &= \frac{1}{N}\mathbb{E}[\|(P^t - N^{-1}\mathbf{1}\mathbf{1}^*)x(0)\|^2] \\
&= \frac{1}{N}\mathbb{E}[\mathrm{tr}((P^t - N^{-1}\mathbf{1}\mathbf{1}^*)x(0)x(0)^*(P^t - N^{-1}\mathbf{1}\mathbf{1}^*))] \\
&= \frac{1}{N}[\mathrm{tr}((P^t - N^{-1}\mathbf{1}\mathbf{1}^*)\mathbb{E}[x(0)x(0)^*](P^t - N^{-1}\mathbf{1}\mathbf{1}^*))] \\
&= \frac{\sigma^2}{N}\,\mathrm{tr}(P^{2t} - N^{-1}\mathbf{1}\mathbf{1}^*).
\end{aligned} \tag{4.4}$$

This formula can be rewritten in terms of the Frobenius norm of a square matrix A, formally defined as $||A||_F := \sqrt{\mathrm{tr}(AA^*)}$. In our case, we have that

$$||P^t - N^{-1}\mathbf{1}\mathbf{1}^*||_F = \sqrt{\mathrm{tr}(P^{2t} - N^{-1}\mathbf{1}\mathbf{1}^*)} = \sqrt{\sum_{i=2}^{N} |\mu_i|^{2t}}.$$

Therefore,

$$N^{-1}\mathbb{E}||P^t x(0) - N^{-1}\mathbf{1}\mathbf{1}^* x(0)||^2 = \frac{\sigma^2}{N}||P^t - N^{-1}\mathbf{1}\mathbf{1}^*||_F^2 = \frac{\sigma^2}{N}\sum_{i=2}^{N} |\mu_i|^{2t}. \tag{4.5}$$

Notice how (4.2) can be directly obtained from (4.5) by simply upper bounding $||P^t - N^{-1}\mathbf{1}\mathbf{1}^*||_F \leq N\rho_2^t$. To illustrate the relation between these two estimates, we propose a few examples.

Example 4.1 (*Complete graph*) If $P = N^{-1}\mathbf{1}\mathbf{1}^*$, we trivially have that $P^t - N^{-1}\mathbf{1}\mathbf{1}^* = 0$ for every t so that $||P^t - N^{-1}\mathbf{1}\mathbf{1}^*||_F = 0$. On the other hand, we also have that $\rho_2 = 0$. The two estimates coincide in this case.

Example 4.2 (*Disconnected graph*) Consider now the simple random walk associated with a disconnected graph consisting of two complete isolated graphs with $N/2$

nodes each (N is assumed to be even):

$$P = \left[\begin{array}{c|c} \frac{2}{N}\mathbf{1}\mathbf{1}^* & 0 \\ \hline 0 & \frac{2}{N}\mathbf{1}\mathbf{1}^* \end{array}\right]. \tag{4.6}$$

The eigenvalues of P are 1 with multiplicity 2 and 0 with multiplicity $N-2$. Therefore,

$$||P^t - N^{-1}\mathbf{1}\mathbf{1}^*||_F = 1, \quad \text{and} \quad \rho_2 = 1.$$

In this case, expressions (4.5) and (4.2) respectively become, for $t \geq 1$,

$$\begin{aligned} N^{-1}\mathbb{E}||P^t x(0) - N^{-1}\mathbf{1}\mathbf{1}^* x(0)||^2 &= \sigma^2 N^{-1}, \\ N^{-1}\mathbb{E}||P^t x(0) - N^{-1}\mathbf{1}\mathbf{1}^* x(0)||^2 &\leq \sigma^2. \end{aligned}$$

Clearly, they are significantly different in terms of N. Notice in particular that the first bound says that, for large N, the mean distance from consensus is small for every value of $t \geq 1$. This difference seems in contrast with the fact that the simple random walk on a disconnected graph does not lead to a consensus. However, notice that the consensus values on the two components are both small for large N with high probability because of the law of large numbers, and therefore, the two consensus values are close to each other.

Perhaps more interestingly, this inconsistency between the two estimates is not limited to disconnected graphs, as we show in the following example, which is in fact a slight modification of the previous one.

Example 4.3 (*Barbell graph*) A *barbell graph*, defined for even N, is a graph composed of two disjoint cliques connected by an edge. The SRW is now

$$\tilde{P} = P + \left[\begin{array}{cc|cc} & & & \\ & -2/N & 2/N & \\ \hline & 2/N & -2/N & \\ & & & \end{array}\right],$$

where P is the SRW (4.6). Matrix $\tilde{P}$ has eigenvalue 1 with multiplicity 1, eigenvalue 0 with multiplicity $N-3$ and two simple eigenvalues $\frac{1}{2} - \frac{2}{N} \pm \frac{1}{2}\sqrt{1 + \frac{8}{N} - \frac{16}{N^2}}$. Here, we rare facing a bottleneck phenomenon due to the single edge connecting the two cliques and this results in a very slow convergence rate $\rho_2(P) = 1 - \frac{8}{N^2} + o(\frac{1}{N^2})$ as $N \to \infty$. Nevertheless, for all $t \geq 1$, it holds

$$||P^t - N^{-1}\mathbf{1}\mathbf{1}^*||_F^2 = \left(\frac{1}{2} - \frac{2}{N} + \frac{1}{2}\sqrt{1 + \frac{8}{N} - \frac{16}{N^2}}\right)^{2t} + \left(\frac{1}{2} - \frac{2}{N} - \frac{1}{2}\sqrt{1 + \frac{8}{N} - \frac{16}{N^2}}\right)^{2t}$$
$$\leq \frac{2}{N}.$$

As in the previous example, the estimation error becomes small already from the first iteration if N is large, but this cannot be seen in the estimation that uses the second eigenvalue ρ_2.

4.2 Rendezvous and Linear-Quadratic Control

The mean convergence rate introduced in Sect. 4.1 is related to the analysis of various other cost functionals. In this section, we consider the consensus dynamics in the context of the rendezvous application, interpreting it as a closed-loop feedback control:

$$x(t+1) = x(t) + u(t) \quad \text{where} \quad u(t) = (P - I)x(t).$$

In this, setting a popular cost functional to measure the control performance of the systems is the quadratic cost defined as $J_{\text{LQ}} := J_x + \epsilon J_u$, where ϵ is a positive weight and

$$J_x := N^{-1} \sum_{t=0}^{\infty} \mathbb{E}||x(t) - N^{-1}\mathbf{1}\mathbf{1}^* x(0)||^2 \tag{4.7}$$

$$J_u := N^{-1} \sum_{t=0}^{\infty} \mathbb{E}||u(t)||^2. \tag{4.8}$$

Cost J_x measures the speed of convergence to consensus, whereas J_u measures the control effort needed to achieve it. The two functionals J_x and J_u can be expressed in terms of the eigenvalues $\{\mu_i\}$ as shown in the following result.

Proposition 4.1 (LQ cost) *If the stochastic matrix P is irreducible aperiodic and symmetric, then*

$$J_x = \frac{\sigma^2}{N} \sum_{t=0}^{\infty} ||P^t - N^{-1}\mathbf{1}\mathbf{1}^*||_F^2 = \frac{\sigma^2}{N} \sum_{i=2}^{N} \frac{1}{1 - \mu_i^2}$$

$$J_u = \frac{\sigma^2}{N} \sum_{t=0}^{\infty} ||P^{t+1} - P^t||_F^2 = \frac{\sigma^2}{N} \sum_{i=2}^{N} \frac{1 - \mu_i}{1 + \mu_i}$$

Proof The expression for J_x is a straightforward consequence of (4.5). Regarding J_u, the first equality comes from a computation analogous to (4.4). The second one instead follows from the observation that the eigenvalues of $(P^{t+1} - P^t)^2$ are given by $\{(\mu_i^{t+1} - \mu_i^t)^2\} = \{\mu_i^{2t}(1-\mu_i)^2\}$. □

The values of J_x and J_u can be effectively estimated or computed in several examples. In general, it is immediate to see that

$$\frac{N-1}{N} \le \frac{J_x}{\sigma^2} \le \frac{1}{1-\rho_2^2},$$

where both bounds are tight (take $P = N^{-1}\mathbf{1}\mathbf{1}^*$). The lower bound implies that J_x is never infinitesimal in the number of nodes, while the upper bound implies that J_x is limited if the second largest eigenvalue of P is bounded away from one. Otherwise, J_x may or may not diverge as N goes to infinity, as shown in the example below.

Let us consider the lazy simple random walk matrix $P = (2d+1)^{-1}(I+A)$ on a d-dimensional torus C_n^d. Eigenvalues can easily be obtained from the eigenvalues of $L(A)$ computed in Example 1.8:

$$\mu_{(h_1,\dots,h_d)} = \frac{1}{2d+1}\left(1 + 2\sum_{i=1}^{d} \cos\frac{2\pi}{n}h_i\right), \quad h_1,\dots,h_d \in \{0,1,\dots,n-1\}$$

Notice now that

$$J_x = \frac{\sigma^2}{n^d} \sum_{(h_1,\dots h_d)\neq 0} \frac{1}{1 - |\mu_{(h_1,\dots,h_d)}|^2}$$

can be interpreted as a Riemann sum of the function $f : [0,1]^d \setminus \{0\} \to \mathbb{R}$ given by

$$f(x) = \frac{\sigma^2}{1 - \left|\frac{1}{2d+1}\left(1 + 2\sum_{i=1}^{d}\cos 2\pi x_i\right)\right|^2}$$

Notice that $f(x)$ presents a singularity in 0: Precisely, we have that $f(x) = \Theta(||x||^{-2})$ for $x \to 0$. This implies that in dimension $d \ge 3$, function f is (absolutely) integrable on $[0,1]^d$. This, combined with the fact that $f(x)$ is monotonic with respect to the each component of x in a neighborhood of 0, implies that

$$\lim_{n\to+\infty} J_x = \int_{[0,1]^d} f(x)\mathrm{d}x < +\infty$$

In particular this shows that on a d-dimensional torus, with $d \ge 3$, J_x is bounded in N. Instead, in dimension 1 and 2, f is no longer integrable and previous argument cannot be applied. Indeed in both cases, J_x turns out to be unbounded in N. In dimension one, an explicit computation shows that (letting $\sigma = 1$):

$$J_x = \frac{1}{N}\sum_{h=1}^{N-1} \frac{1}{1-\frac{1}{9}(1+2\cos\left(\frac{2\pi}{N}h\right))^2} \geq \frac{N^{-1}9/4}{2-\cos\left(\frac{2\pi}{N}\right)-\cos^2\left(\frac{2\pi}{N}\right)} \geq \frac{3}{8\pi^2}N. \tag{4.9}$$

A more detailed analysis including the two-dimensional case is provided in Exercise 5.11, by using the tools developed in that chapter.

On the contrary, J_u shows better scaling properties: for the lazy SRW on the cycle,

$$J_u = \frac{1}{N}\sum_{h=1}^{N-1} \frac{1-\cos\left(\frac{2\pi}{N}h\right)}{2+\cos\left(\frac{2\pi}{N}h\right)},$$

which is clearly bounded in N. By interpreting it as a Riemann sum one can see that, more precisely,

$$\lim_{n\to+\infty} J_u = \int_0^1 \frac{1-\cos(2\pi x)}{2+\cos(2\pi x)}\mathrm{d}x = \sqrt{3}-1.$$

Other examples are given in Exercise 4.10. More generally, J_u can be shown to be bounded under weak assumptions. To this goal, we recall a well-known property of the spectrum of a matrix.

Lemma 4.1 (Gershgorin) *Let A be an $n \times n$ matrix. Then,*

$$\mathrm{spec}(A) \subset \bigcup_{i\in\{1,\ldots,n\}} \{z \in \mathbb{C} : |z - a_{ii}| \leq \sum_{j\neq i} |a_{ij}|\}.$$

An immediate application of this lemma yields the following result.

Proposition 4.2 (Boundedness of J_u) *Let P be such that $P_{vv} > 0$ for all $v \in V$, and denote $\alpha = \min_v P_{vv}$. Then,*

$$J_u \leq \frac{1-\alpha}{\alpha}.$$

4.3 Robustness Against Noise

In this section, we analyze the behavior and performance of consensus algorithms under the presence of noise in the dynamics. As we will see, cost functionals similar to those introduced above naturally come up in this case. Noise is unavoidable in many applications. Instances can be imprecisions in the motion of robots in the rendezvous problem or quantization errors in digital transmissions among the nodes of the network. In this section, we analyze the effects of noise in several models where the consensus dynamics is perturbed in different ways. We recall the standing assumption that P is a symmetric stochastic matrix.

We start considering the case when noise enters additively in the update equation (this can be a model for the robots motion error):

$$x(t+1) = Px(t) + n(t) \tag{4.10}$$

We assume the $n_v(t)$ to be independent random variables with mean 0 and variance σ^2. Notice first that the mean value is governed by $\mathbb{E}[x(t+1)] = P\mathbb{E}[x(t)]$ so that, if P is irreducible and aperiodic, we have convergence to the average consensus: $\mathbb{E}[x(t)] \to N^{-1}\mathbf{1}\mathbf{1}^* x(0)$ (here, $x(0)$ is seen as deterministic). If we define $m(t) = N^{-1}\sum_v x_v(t)$ and $\mu(t) = N^{-1}\sum_v n_v(t)$, Eq. (4.10) implies that

$$m(t+1) = m(t) + \mu(t). \tag{4.11}$$

Consequently,

$$m(t) = m(0) + \sum_{s=0}^{t-1} \mu(s)$$

Notice that each $\mu(t)$ is a r.v. with mean 0 and variance σ^2/N. Therefore, we can conclude that $m(t)$ is a process with

$$\mathbb{E}[m(t)] = m(0), \quad \mathrm{Var}[m(t)] = \sigma^2 \frac{t}{N}. \tag{4.12}$$

This shows how noise accumulates into the linear dynamics (essentially because of its marginally stable structure) and creates such unbounded effects on the average dynamics. A similar phenomenon takes place if we measure the distance of the process from a consensus point. Consider indeed the following functional

$$J_{\text{noise}} = \frac{1}{N} \lim_{t\to+\infty} \mathbb{E}||x(t) - N^{-1}\mathbf{1}\mathbf{1}^* x(t)||^2 \tag{4.13}$$

Remarkably, J_{noise} coincides with the functional J_x introduced to describe the LQ cost functional.

Proposition 4.3 (Noise cost) *Suppose that P is a symmetric irreducible and aperiodic stochastic matrix. Then,* $J_{\text{noise}} = J_x = \frac{\sigma^2}{N}\sum_{t=0}^{+\infty} ||P^{2t} - N^{-1}\mathbf{1}\mathbf{1}^*||_F^2$.

Proof It follows from (4.10) that, for every time t, it holds

$$x(t) = P^t x(0) + \sum_{s=0}^{t-1} P^s n(t-s-1)$$

which yields

$$\begin{aligned}\mathbb{E}||x(t)- N^{-1}\mathbf{1}\mathbf{1}^*\ x(t)||^2 &= \mathbb{E}||(P^t - N^{-1}\mathbf{1}\mathbf{1}^*)x(0)||^2 \\ &+ \sum_{s=0}^{t-1}\sum_{s'=0}^{t-1} \mathbb{E}[(P^s - N^{-1}\mathbf{1}\mathbf{1}^*)n(t-s-1)]^*[(P^{s'} - N^{-1}\mathbf{1}\mathbf{1}^*)n(t-s'-1)] \\ &+ 2\sum_{s=0}^{t-1} \mathbb{E}[((P^t - N^{-1}\mathbf{1}\mathbf{1}^*)x(0))^*(P^s - N^{-1}\mathbf{1}\mathbf{1}^*)n(t-s-1)]\end{aligned}$$

Now, the first term converges to 0, when $t \to +\infty$, because of the assumptions made on P. The third term is 0 because all noises are zero mean. Finally, the second term can be rewritten as

$$\sigma^2 \sum_{s=0}^{t-1} \mathrm{tr}[P^{2s} - N^{-1}\mathbf{1}\mathbf{1}^*] = \sigma^2 \sum_{s=0}^{t-1} ||P^s - N^{-1}\mathbf{1}\mathbf{1}^*||_F^2$$

where we have used the independence assumption on the noises. By taking the limit $t \to +\infty$, we obtain the result. □

The example in Eq. (4.9) implies then that J_{noise} is in general unbounded in N for large-scale graphs. Moreover, Eq. (4.12) shows that the variance of the average value diverges with time. These considerations demonstrate that consensus dynamics is sensitive to additive noise: This sensitivity is stronger for matrices P that have eigenvalues closer to the unit circle. Actually, when the dispersion of the initial condition is small with respect to the variance of the noise and to the number of nodes N, running a consensus algorithm may even be detrimental in terms of $\mathbb{E}||x(t) - N^{-1}\mathbf{1}\mathbf{1}^*x(t)||^2$. These circumstances are explored in Exercise 4.8. Even outside such extreme cases, sensitivity to noise can be a problem in practical applications and some countermeasures against noise have thus been proposed. A useful idea is replacing the time-invariant averaging dynamics with a time-varying version that smooths out the effects of noise by employing a "decreasing gain" strategy. We do not cover these more refined algorithms here, but some literature pointers are given at the end of this chapter.

4.4 Robustness Against Quantization Errors

If we assume that communication among units takes place through digital channels, then the communicated states will be affected by rounding (or quantization) errors. These errors are unavoidable because the state is real-valued, whereas the communicated values are discrete. These errors, which depend on the state and on the quantization rule, can be modeled as independent stochastic noises with zero mean and with variance σ^2 determined by the precision of the approximation. This modeling leads to consider a dynamics like

$$x(t+1) = P\big(x(t) + n(t)\big). \tag{4.14}$$

It is easy to realize that this dynamics suffers from the same drawbacks as dynamics (4.10). In particular, the average process $m(t)$ is governed by the same relation (4.11), and consequently, the same conclusions on the moments (4.12) can be drawn. Moreover, the functional defined as in (4.13) can be evaluated similarly to Proposition 4.3, as detailed in Exercise 4.7.

Notice, however, that in this case one can run, instead of (4.14), the alternative averaging dynamics

$$x(t+1) = P\big(x(t) + n(t)\big) - n(t). \tag{4.15}$$

In this dynamics, we subtract the noise $n(t)$: This operation is feasible as it is realistic to assume that each node v knows $n_v(t)$, that is, the quantization error affecting its own value. This dynamics is chosen with the purpose of reducing the effect of the noise. Indeed, differently from (4.10), dynamics (4.15) deterministically preserves the average of the initial condition:

$$N^{-1}\mathbf{1}\mathbf{1}^* x(t+1) = N^{-1}\mathbf{1}\mathbf{1}^* x(t) + N^{-1}\mathbf{1}\mathbf{1}^* n(t) - N^{-1}\mathbf{1}\mathbf{1}^* n(t) = N^{-1}\mathbf{1}\mathbf{1}^* x(t).$$

The asymptotical dispersion around the average can be evaluated by using the functional

$$J_{\mathrm{q}} = \lim_{t\to+\infty} \mathbb{E}||x(t) - N^{-1}\mathbf{1}\mathbf{1}^* x(t)||^2,$$

which is formally defined as (4.13) but with the understanding that here $x(t)$ follows (4.15). Perhaps surprisingly, J_{q} coincides with the functional J_u introduced to describe the LQ cost functional, as the reader can verify as an exercise.

Proposition 4.4 (Quantization cost) *Suppose that P is a symmetric irreducible and aperiodic stochastic matrix. Then,* $J_{\mathrm{q}} = J_u = \frac{\sigma^2}{N}\sum_{t=0}^{\infty} ||P^{t+1} - P^t||_F^2$.

By recalling the results of Sect. 4.2, the reader can see that the effect of noise is largely reduced in (4.15), compared to (4.14).

4.5 Distributed Inference

An important application of consensus is solving, in a distributed fashion, network inference problems. Below we discuss some basic examples and we show how the analysis of performance, also in this case, leads to functionals similar to those considered before.

Assume that each node $v \in V$ takes a measurement of the same unknown scalar quantity θ. Each of these measurements, denoted by y_v, is affected by an (additive) measurement error n_v. Namely, $y_v = \theta + n_v$. The goal of each node is to estimate θ. The node v by itself could only estimate θ by the taken measurement y_v, whereas if it was possible to gather the measurements from all nodes, more efficient estimation could be performed. If we assume the measurement errors to be independent random variables with zero mean and variance σ^2, then the average $\hat{\theta} = \frac{1}{N}\sum_v y_v$ is the optimal estimator for θ and has estimation error $\mathbb{E}||\hat{\theta} - \theta||^2 = N^{-1}\sigma^2$. Conveniently, the network may collectively compute $\hat{\theta}$ by simply using the dynamics (4.1) with $x_v(0) = y_v$ for all nodes v. In this context, it is natural to define the time-dependent *estimation error* as

$$J_e(t) = \frac{1}{N}\mathbb{E}[\|x(t) - \theta\mathbf{1}\|^2].$$

If we denote by $n \in \mathbb{R}^V$ the random vector collecting all noises and we repeat the computation as in (4.4), we see that $J_e(t)$ can be rewritten as

$$J_e(t) = \frac{1}{N}\mathbb{E}[\|P^t n\|^2] = \sigma^2 N^{-1}||P^t||_F^2 = \sigma^2 N^{-1}\sum_{i=1}^{N}\mu_i^{2t} \tag{4.16}$$

Notice the difference with respect to (4.5), where we had the Frobenius norm of $P^t - N^{-1}\mathbf{1}\mathbf{1}^*$. Indeed, differently from (4.5) that converges to 0 for $t \to +\infty$, we here have $J_e(t) \to \frac{\sigma^2}{N}$ as $t \to \infty$. The asymptotic error is due to the intrinsic mean estimation error. The consensus algorithm (4.1) can also be used to solve more general inference problems, in which the measurements errors can have different variances σ_v^2. Again, the goal of each node is to estimate θ. The node v by itself could only estimate θ by the taken measurement y_v, whereas if it was possible to gather the measurements from all nodes, more efficient estimation could be performed. In the latter case, the best least squares estimator, defined as

$$\widehat{\theta} := \text{argmin}_\theta \sum_v \frac{(y_v - \theta)^2}{\sigma_v^2},$$

can be computed as

$$\widehat{\theta} = \left(\sum_{w\in V}\frac{1}{\sigma_w^2}\right)^{-1}\sum_{v\in V}\frac{y_v}{\sigma_v^2}.$$

This estimator, which is a Maximum Likelihood estimator when the measurement noises are Gaussian (see Exercise 4.1), simply becomes the average of the measurements when all variances are equal. Clearly, to compute such an estimator, one needs to gather all the measurements y_v's. However, rewriting it as

$$\widehat{\theta} = \left(\frac{1}{N}\sum_{w\in V}\frac{1}{\sigma_w^2}\right)^{-1}\frac{1}{N}\sum_{v\in V}\frac{y_v}{\sigma_v^2},$$

one can notice that it is the ratio between two arithmetic means. Then, consensus algorithms can be naturally applied to approximate it, provided each node knows the variance of its own measurement error. Consider two consensus algorithms built on the matrix P and running in parallel:

$$x^{(1)}(t+1) = Px^{(1)}(t), \quad x_v^{(1)}(0) = \frac{y_v}{\sigma_v^2} \quad \forall v \in V$$

$$x^{(2)}(t+1) = Px^{(2)}(t), \quad x_v^{(2)}(0) = \frac{1}{\sigma_v^2} \quad \forall v \in V$$

and define $\widehat{\theta}_v(t) = x_v^{(1)}(t)/x_v^{(2)}(t)$. We know from the results of Chap. 2 (see also Exercise 4.2) that

$$\lim_{t \to +\infty} \widehat{\theta}(t)_v = \widehat{\theta} \quad \forall v \in V.$$

Hence, the estimator can be computed by running two consensus algorithms and computing the ratio of their states. In the exercises, we propose a few adaptations and variations of the above procedure.

Exercises

Exercise 4.1 (*Maximum Likelihood estimator*) For all $v \in V$, let $y_v = \theta + n_v$ and assume that each n_v is a measurement error to be independent Gaussian random variables with zero mean and variance σ_v^2. Consider the density distribution of y_v given that the unknown quantity is θ

$$f(y_v \,|\, \theta) = \frac{1}{\sqrt{2\pi\sigma_v^2}} e^{\frac{(y_v-\theta)^2}{2\sigma_v^2}}$$

and the global density of the vector $y \in \mathbb{R}^V$

$$f(y \,|\, \theta) = \prod_v f(y_v \,|\, \theta) = \prod_v \frac{1}{\sqrt{2\pi\sigma_v^2}} e^{\sum_v \frac{(y_v-\theta)^2}{2\sigma_v^2}}$$

The ML estimator is defined to be $\widehat{\theta}^{ML} := \operatorname{argmax}_{\theta \in \mathbb{R}} f(y \,|\, \theta)$. Verify that

$$\widehat{\theta}^{ML} = \left(\sum_{w \in V} \frac{1}{\sigma_w^2} \right)^{-1} \sum_{v \in V} \frac{y_v}{\sigma_v^2}$$

Exercise 4.2 (*Consensus ratio*) Let $P \in \mathbb{R}^{V \times V}$ be a stochastic irreducible aperiodic matrix. Consider the dynamics

$$\begin{cases} x(t+1) = P\,x(t) & x(0) \in \mathbb{R}^V \\ y(t+1) = P\,y(t) & y(0) \in \mathbb{R}^V,\ y_v(0) > 0\ \forall v \in V. \end{cases}$$

Let $z(t) = \frac{x(t)}{y(t)}$. Determine $z(\infty) := \lim_{t \to +\infty} z(t)$.

Exercise 4.3 (*Average consensus with nondoubly stochastic matrices*) Let $P \in \mathbb{R}^{V \times V}$ be a stochastic irreducible aperiodic matrix. Consider the dynamics

$$\begin{cases} x(t+1) = P^* x(t) & x(0) \in \mathbb{R}^V \\ y(t+1) = P^* y(t) & y(0) \in \mathbb{R}^V,\ y_v(0) > 0\ \forall v \in V. \end{cases}$$

(i) Let $z(t) = \frac{x(t)}{y(t)}$. Determine $z(\infty) := \lim_{t \to +\infty} z(t)$.
(ii) Given a vector $\bar{x} \in \mathbb{R}^V$, choose $x(0)$ and $y(0)$ in such way that $z(\infty) = \mathbf{1}N^{-1}\mathbf{1}^*\bar{x}$.

Exercise 4.4 (*Weighted averages*) Using a doubly stochastic matrix P, design a consensus-based algorithm to compute any weighted average of values known to the nodes.

Exercise 4.5 (*Number of nodes*) Design a consensus-based algorithm for a network to compute the number of its nodes.

Exercise 4.6 (*Least Squares Regression*) We want to estimate a function $y = f(x)$ from a noisy data set $\{(x_v, y_v)\}_{v \in V}$ collected by the nodes.We parameterize $f(\cdot)$ according to a basis of functions $\{g_j(\cdot)\}_{j \in J}$, where J a suitable index set, so that $f_\theta(x) = \sum_{j \in J} \theta_j g_j(x)$. The basis functions are known, and the $|J|$-dimensional vector θ is to be determined. We want to compute (distributely) the best estimate of θ in a least squares sense. Provided we define $G \in \mathbb{R}^{J \times V}$ to be a matrix such that $G_{jv} = g_j(x_v)$, the optimal estimator is defined as

$$\hat{\theta} = \mathrm{argmin}_\theta \|y - G\theta\|^2.$$

(i) Verify that $\hat{\theta} = (G^*G)^{-1}G^*y$.

Let g^v denote a column of G and define two consensus algorithms with initial conditions

$$\begin{aligned} z_v^{(1)}(0) &= g^v (g^v)^* \quad \forall v \in V \\ z_v^{(2)}(0) &= g^v y_v \quad \forall v \in V. \end{aligned}$$

Note that $z_v^{(1)} \in \mathbb{R}^{J \times J}$ and $z_v^{(2)} \in \mathbb{R}^J$. Since the states are non scalar, the update is performed independently on each component.

(ii) Remark that $\hat{\theta} = (\sum_v g^v (g_v)^*)^{-1} \sum_v g^v y_v$, and deduce that

$$\lim_{t \to +\infty} (z_v^{(1)}(t))^{-1} z_v^{(2)}(t) = \hat{\theta}.$$

Exercise 4.7 (*Communication noise*) Consider a symmetric stochastic matrix P on V and the process $x(t)$ taking values in $\mathbb{R}^V$ and governed by equation

$$x_v(t+1) = \sum_{w \in V} P_{vw} \big(x_w(t) + n_{vw}(t) \big)$$

where $\{n_{vw})(t)\}$ is a family of independent 0 mean, σ^2 variance random variables. In this model, we assume that noises are independently generated in any pairwise transmission between units. For such process, study the behavior of the corresponding average process $m(t)$ and find an expression for the functional as defined in (4.13) in terms of the eigenvalues of P.

Exercise 4.8 (*Noise effects at finite times*) We define the following time-dependent version of the noise cost (4.13)

$$J_{\text{noise}}(t) = \frac{1}{N}\mathbb{E}||x(t) - N^{-1}\mathbf{1}\mathbf{1}^* x(t)||^2, \tag{4.17}$$

assuming that the noise components are iid random variables with zero mean and variance σ_n^2, while initial conditions are iid random variables with zero mean and variance σ_x^2 independent of the noise.

(i) By proceeding as in the proof of Proposition 4.3, show that

$$J_{\text{noise}}(t) = \frac{1}{N}\sum_{i=2}^{N}\frac{\sigma_n^2}{1-|\mu_i|^2} + \frac{1}{N}\sum_{i=2}^{N}\left(\sigma_x^2 - \frac{\sigma_n^2}{1-|\mu_i|^2}\right)|\mu_i|^{2t} \tag{4.18}$$

(ii) Observe that if σ_x is small enough (for instance if $\sigma_x < \sigma_n$), then $J_{\text{noise}}(t)$ is increasing with time.

(iii) Verify that the second term of (4.18) is upper bounded by $\frac{1}{N}\left(N\sigma_x^2 - \frac{\sigma_n^2}{1-\rho_2^2}\right)\rho_2^{2t}$

(iv) Assume that P is the lazy simple random walk matrix P on the cycle graph C_N (cf. Exercise 2.12). Verify that if $N > \frac{8\pi}{9}\frac{\sigma_x^2}{\sigma_n^2}$, then $J_{\text{noise}}(t)$ is increasing with time.

Exercise 4.9 (*Normal update matrix*) Reconsider system (4.1) with the assumption that the irreducible and aperiodic matrix P is doubly stochastic and normal (but not necessarily symmetric). Show that

$$J_e(t) = \frac{1}{N}\sum_i |\mu_i|^{2t} \quad J_x = \frac{1}{N}\sum_{i>1}\frac{1}{1-|\mu_i|^2} \quad J_u = \frac{1}{N}\sum_{i>1}\frac{|1-\mu_i|^2}{1-|\mu_i|^2}.$$

Exercise 4.10 (J_u *cost* [11]) Consider the cost J_u defined in Sect. 4.2.

(i) Let $G = \mathbf{C}_N$ be a directed cycle graph and $P = \text{circ}(1/2, 1/2, 0, \ldots, 0)$. Then, $J_u = 1 - \frac{1}{N}$.

(ii) Let $G = \mathbf{C}_n^d$ be a directed d-dimensional torus graph and $P = \frac{1}{d+1}(I + A_G)$. Then, $J_u = 1 - \frac{1}{n^d}$.

(iii) Let G be a d-dimensional hypercube and $P = \frac{1}{d+1}(I + A_G)$. Then, $J_u = 1 - \frac{1}{2^d}$.

Exercise 4.11 (J_e *cost on toroidal grids* [14]) Let $G = C_n^d$ be a d-dimensional torus graph, $P = \frac{1}{2d+1}(I + A_G)$ and $N = n^d$.

(i) Verify that

$$c_1 \max\{\frac{1}{N}, \frac{1}{t^{d/2}}\} \leq J_e(t) \leq c_2 \max\{\frac{1}{N}, \frac{1}{t^{d/2}}\} \quad \text{for some positive } c_1, c_2.$$

(ii) Estimate the time needed to achieve the best precision in the estimation for a given N.

Exercise 4.12 (*Size optimization in distributed estimation* [14]) Based on the functional $J_e(t)$, we consider the problem of optimizing the size of the graph in a specific family of consensus matrices. Let A_n be the adjacency matrix of cycle graph C_n of order n. Let $P_n = \frac{1}{3}(I + A_n)$.

(i) Verify that
  (a) $J(P_n, t)$ is nonincreasing in n;
  (b) $J(P_n, t)$ is nonincreasing in t;
  (c) $J(P_n, t) = J(P_{2t+1}, t)$ for all $n \geq 2t + 1$.

(ii) Discuss the results above from the point of view of design, having the goal of efficient estimation of a parameter which is known via noisy measurements. Is there a "best" size of the network, if the available time for computation is limited?

Bibliographical Notes

An introduction to the costs considered in this chapter is available in [13], which also provides useful pointers to the literature. These costs have been explicitly computed using the eigenvalues of P, thanks to the assumption of symmetry: However, one can generalize this analysis to normal matrices (see Exercise 4.9 and [7]) and to reversible matrices (see Exercise 5.10).

Specific references can be given for the different functionals considered. For instance, a thorough analysis of J_e on geometric graphs is given by [14]. Cost J_{noise} has been defined in the seminal paper [25] and later extensively studied with different interpretations and variations [17, 21]. Paper [11] has proposed dynamics (4.14) to cope with quantization errors and has studied cost J_q. The statistical assumptions on the quantization errors can either be rigorously justified for certain randomized quantizers (for instance, quantizers with "dithering" [1]) or taken as a useful approximation for deterministic quantizers. Actually, several researchers have looked at quantization in the context of averaging algorithms, starting with [19]: A selection of the papers that are most closely related to our perspective includes [3, 6, 9, 18, 20, 22].

As we mentioned, the effects of noise entering the averaging system can be mitigated by using properly designed decreasing gains. This adaptation typically results in systems that almost surely converge to consensus, but such that convergence is not exponentially fast. Their analysis can benefit from the so-called stochastic approximation techniques [2]. Many papers have taken this approach to ensure robustness, including [5, 8, 16, 23, 24].

The application of distributed parameter estimation has also been presented in this chapter. In the literature, this problem has been extended in various directions, including least squares regression (see Exercise 4.6), distributed Kalman filtering [4], and estimation of parameters that are vector-valued and distributed over the nodes (see Sect. 5.4). The treatment given in this chapter assumes that each node knows the variance of its own measurement error. If this is not the case, a more complex algorithm is needed in order to estimate these quantities as well. For instance, paper [10] looks at a special case of this problem, where nodes are divided into two classes, having respectively small and large variance, and must identify to which class they belong to. The issue of parameter estimation can also be interpreted in the context of social networks. Empirical and theoretical evidences have shown that aggregate opinions may provide a good estimate of unknown quantities: Such phenomenon has been referred to in the literature as the wisdom of crowds [12, 15].

References

1. Aysal, T.C., Coates, M.J., Rabbat, M.G.: Distributed average consensus with dithered quantization. IEEE Trans. Signal Process. **56**(10), 4905–4918 (2008)
2. Borkar, V.S.: Stochastic Approximation: A Dynamical Systems Viewpoint. Cambridge University Press (2008)
3. Carli, R., Bullo, F., Zampieri, S.: Quantized average consensus via dynamic coding/decoding schemes. Int. J. Robust Nonlinear Control **20**(2), 156–175 (2010)
4. Carli, R., Chiuso, A., Schenato, L., Zampieri, S.: Distributed Kalman filtering based on consensus strategies. IEEE J. Sel. Areas Commun. **26**(4), 622–633 (2008)
5. Carli, R., Como, G., Frasca, P., Garin, F.: Distributed averaging on digital erasure networks. Automatica **47**(1), 115–121 (2011)
6. Carli, R., Fagnani, F., Speranzon, A., Zampieri, S.: Communication constraints in the average consensus problem. Automatica **44**(3), 671–684 (2008)
7. Carli, R., Garin, F., Zampieri, S.: Quadratic indices for the analysis of consensus algorithms. In: Information Theory and Applications Workshop, San Diego, CA, pp. 96–104, February 2009
8. De Persis, C., Frasca, P.: Robust self-triggered coordination with ternary controllers. IEEE Trans. Autom. Control **58**(12), 3024–3038 (2013)
9. El Chamie, M., Liu, J., Başar, T.: Design and analysis of distributed averaging with quantized communication. IEEE Trans. Autom. Control **61**(12), 3870–3884 (2016)
10. Fagnani, F., Fosson, S.M., Ravazzi, C.: A distributed classification/estimation algorithm for sensor networks. SIAM J. Control Optim. **52**(1), 189–218 (2014)
11. Frasca, P., Carli, R., Fagnani, F., Zampieri, S.: Average consensus on networks with quantized communication. Int. J. Robust Nonlinear Control **19**(16), 1787–1816 (2009)
12. Galton, F.: Vox populi. Nature **75**(1949), 450–451 (1907)
13. Garin, F., Schenato, L.: A survey on distributed estimation and control applications using linear consensus algorithms. In: Bemporad, A., Heemels, M., Johansson, M. (eds.) Networked Control Systems. Lecture Notes in Control and Information Sciences. Springer (2010)
14. Garin, F., Zampieri, S.: Mean square performance of consensus-based distributed estimation over regular geometric graphs. SIAM J. Control Optim. **50**(1), 306–333 (2012)

15. Golub, B., Jackson, M.O.: Naïve learning in social networks and the hegsel of crowds. American Econ. J. Microecon. **2**(1), 112–149 (2010)
16. Huang, M., Manton, J.H.: Coordination and consensus of networked agents with noisy measurements: stochastic algorithms and asymptotic behavior. SIAM J. Control Optim. **48**(1), 134–161 (2009)
17. Jadbabaie, A., Olshevsky, A.: On performance of consensus protocols subject to noise: role of hitting times and network structure. In: IEEE Conference on Decision and Control, Las Vegas, NV, pp. 179–184, December 2016
18. Kar, S., Moura, J.M.F.: Distributed consensus algorithms in sensor networks: Quantized data and random link failures. IEEE Trans. Signal Process. **58**(3), 1383–1400 (2010)
19. Kashyap, A., Başar, T., Srikant, R.: Quantized consensus. Automatica **43**(7), 1192–1203 (2007)
20. Li, T., Fu, M., Xie, L., Zhang, J.-F.: Distributed consensus with limited communication data rate. IEEE Trans. Autom. Control **56**(2), 279–292 (2011)
21. Lovisari, E., Garin, F., Zampieri, S.: Resistance-based performance analysis of the consensus algorithm over geometric graphs. SIAM J. Control Optim. **51**(5), 3918–3945 (2013)
22. Nedic, A., Olshevsky, A., Ozdaglar, A., Tsitsiklis, J.N.: On distributed averaging algorithms and quantization effects. IEEE Trans. Autom. Control **54**(11), 2506–2517 (2009)
23. Olshevsky, A.: Consensus with ternary messages. SIAM J. Control Optim. **52**(2), 987–1009 (2014)
24. Rajagopal, R., Wainwright, M.J.: Network-based consensus averaging with general noisy channels. IEEE Trans. Signal Process. **59**(1), 373–385 (2011)
25. Xiao, L., Boyd, S., Kim, S.-J.: Distributed average consensus with least-mean-square deviation. J. Parallel Distrib. Comput. **67**(1), 33–46 (2007)

Chapter 5
Averaging with Exogenous Inputs and Electrical Networks

Abstract The dynamical models analyzed so far, with the exception of the noisy consensus models treated in Chap. 4, are autonomous systems with no input signals: Information enters the system only through the initial condition. Instead, there are a variety of different situations where it is natural to consider consensus models driven by exogenous input signals, including opinion dynamics in the presence of stubborn agents that do not modify their opinion, rendezvous problems with leader robots, and estimation algorithms based on pairwise measurements. A very useful tool to analyze these models is thinking of the graph as an electrical circuit with the exogenous signals interpreted as input currents or as nodes kept at a fixed voltage. In this chapter, we will first review the basic theory of electrical networks and their classical connection with reversible stochastic matrices: Sect. 5.1 concentrates on Green matrices and harmonic functions, while Sect. 5.2 is devoted to effective resistances. Afterward, we apply these tools to averaging dynamics with stubborn agents in Sect. 5.3 and to the problem of estimation from relative measurements in Sect. 5.4.

5.1 Electrical Networks and Harmonic Functions

There is a fundamental connection between reversible stochastic matrices, presented in Sect. 2.5, and electrical circuits: This connection sheds light on some of the concepts touched so far and, meanwhile, offers computational tools for new problems.

We start from a symmetric strongly connected graph $G = (V, E)$ with $|V| = N$ and a symmetric nonnegative matrix $C \in \mathbb{R}^{V \times V}$ called *conductance matrix* such that $\mathcal{G}_C = G$. We know that from C we can canonically construct a reversible stochastic matrix $P = D_{C1}^{-1}C$. We now interpret G as an electrical circuit where edge (u, v) has electrical conductance $C_{uv} = C_{vu}$. We will refer to (G, C) as to an *electrical network*. Consider now a vector $\iota \in \mathbb{R}^V$ such that $\iota^*\mathbf{1} = 0$: We interpret ι_v as the *input current* injected at node v (if negative being an outgoing current). To the electrical network (G, C) and the input current ι, we can associate two functions $W \in \mathbb{R}^V$ (called the *voltage*) and $\phi \in \mathbb{R}^E$ (called the *current flow*) such that the following relations are satisfied

© Springer International Publishing AG 2018
F. Fagnani and P. Frasca, *Introduction to Averaging Dynamics over Networks*, Lecture Notes in Control and Information Sciences 472,
https://doi.org/10.1007/978-3-319-68022-4_5

$$\begin{cases} \sum_{v \in N_u} \phi_{uv} = \iota_u \ \forall u \in V \\ \phi_{uv} = C_{uv}(W_u - W_v) \ \forall (u, v) \in E. \end{cases} \tag{5.1}$$

The first relation is usually known as Kirchoff's law (sum of currents outgoing node u along the edges equals the incoming input current), while the second one is Ohm's law. The existence of a solution (W, ϕ) will follow by our considerations below, as well as uniqueness up to addition to W of multiples of $\mathbf{1}$. Notice that because of Ohm's law, it follows that $\phi_{uv} = -\phi_{vu}$ for all $(u, v) \in E$.

To the aim of rewriting in a more compact form relations (5.1), it is convenient to introduce some additional concepts. Denote by $\bar{E}$ the set of *undirected* edges of G: Namely $\bar{E}$ consists of those subsets $\{u, v\}$ of cardinality 2 such that $(u, v) \in E$ (possible self-loops present in G are disregarded in the construction of $\bar{E}$). An *incidence matrix* on G is any matrix $B \in \{0, +1, -1\}^{\bar{E} \times V}$ such that $B\mathbf{1} = 0$ and $B_{eu} \neq 0$ iff $u \in e$. It is immediate to see that given $e = \{u, v\}$, the e-th row of B has all zeroes except B_{eu} and B_{ev}: Necessarily one of them will be $+1$ and the other one -1 and this will be interpreted as choosing a direction in e from the node corresponding to $+1$ to the one corresponding to -1. Define $D_C \in \mathbb{R}^{\bar{E} \times \bar{E}}$ to be the diagonal matrix such that $(D_C)_{ee} = C_{uv} = C_{vu}$ if $e = \{u, v\}$. Observe that, for every $u \in V$,

$$(B^* D_C B)_{uu} = \sum_{e \in \bar{E}} (D_C)_{ee} B_{eu}^2 = (C\mathbf{1})_u - C_{uu}$$

while, if $u \neq v$,

$$(B^* D_C B)_{uv} = \sum_{e \in \bar{E}} B_{eu} (D_C)_{ee} B_{ev} = -C_{uv}$$

In other terms

$$B^* D_C B = D_{C\mathbf{1}} - C = L(C) .$$

In the special case when $C = A_G$ (the adjacency matrix of G), we thus obtain $B^* B = L_G$. Finally, define $\bar{\phi} \in \mathbb{R}^{\bar{E}}$ such that $\bar{\phi}_e = \phi_{uv} B_{eu}$ if $e = \{u, v\}$: According to this definition, $\bar{\phi}_e$ is the current flowing in the edge e, with the positive sign if flow is happening in the same direction of the conventional direction chosen on e by B. We can now rewrite relations (5.1) as

$$\begin{cases} B^* \bar{\phi} = \iota \\ D_C B W = \bar{\phi}. \end{cases} \tag{5.2}$$

These two equations together lead to the following equation for W:

$$L(C) W = \iota. \tag{5.3}$$

Recall that $L(C)$ is a symmetric matrix with rank $L(C) = N - 1$ and $L(C)\mathbf{1} = 0$ (see Chap. 1 for details). It thus admits the spectral representation

$$L(C) = \sum_{i\geq 2} \lambda_i x_i x_i^*,$$

where $0 = \lambda_1 < \lambda_2 \leq \cdots \leq \lambda_N$ are the nonzero eigenvalues with corresponding orthonormal eigenvectors $N^{-1/2}\mathbf{1}, x_1, \ldots, x_N$. The matrix

$$Z_C = \sum_{i\geq 2} \lambda_i^{-1} x_i x_i^*$$

is said to be the *Green matrix* associated with C. It has the properties

$$Z_C L(C) = L(C) Z_C = I - N^{-1}\mathbf{1}\mathbf{1}^*, \quad Z_C\mathbf{1} = 0. \tag{5.4}$$

If we consider $W = Z_C\iota$, using the property $\iota^*\mathbf{1} = 0$, we obtain $L(C)W = L(C)Z_C\iota = \iota$: Then, W solves (5.3). Notice that it also satisfies the relation $\mathbf{1}^*W = 0$ and that any other function $W + c\mathbf{1}$ also satisfies (5.3). Since the rank of $L(C)$ is $N - 1$, these are all the possible solutions. All pairs solving (5.3) are thus

$$W = Z_C\iota + c\mathbf{1}, \quad \bar{\phi} = CDBZ_C\iota.$$

We now give some insightful examples of computations of voltages.

Example 5.1 (*Line graph*) Consider the symmetric line graph $G = L_{N+1}$ (with vertex set $V = \{1, \ldots, N+1\}$) and with conductance matrix $C \in \mathbb{R}^{V\times V}$. Let $\iota \in \mathbb{R}^V$ be such that $-\iota_1 = 1 = \iota_{N+1}$ while $\iota_k = 0$ for all $k = 2, \ldots, N$. Using Kirchoff's law and a simple inductive argument, it follows that the current flow $\phi \in \mathbb{R}^{E\times E}$ is given by $\phi_{k,k+1} = -1$ for all $k = 1, \ldots, N$. Ohm's law then yields $W_{k+1} - W_k = C_{k,k+1}$ for all k. This yields $W_k - W_0 = \sum_{j=0}^{k-1} C_{j,j+1}$ In the special case when $C = A_G$ (all edges have conductance equal to 1), we obtain $W_k - W_0 = k$.

Example 5.2 (*Leaves and branches*) Let $G = (V, E)$ be a symmetric graph, C a conductance matrix and $\iota \in \mathbb{R}^V$ an input current (with $\iota^*\mathbf{1} = 0$). Let $v \in V$ be such that $\iota_v = 0$ and $d_v = 1$. Consider the longest path in G, $v_1 = v, v_2, \ldots, v_n$ with the property that $\iota_{v_k} = 0$ and $d_{v_k} = 2$ for all $k = 2, \ldots, n-1$. Since $\iota_{v_1} = 0$, Kirchoff law implies that no current can flow in the edge (v_1, v_2) and a simple inductive argument yields that the same happens in all edges (v_{k-1}, v_k) for $k = 3, \ldots n$. Ohm's law then implies that $W_{v_1} = \cdots = W_{v_n}$.

Example 5.3 (*Toroidal grid*) Consider the toroidal 2-grid $G = C_n \times C_n$ with unitary conductances ($C = A_G$). We know from Example 1.7 that its Laplace matrix $L(G) = L(C)$ has eigenvalues

$$\lambda_{(h,k)} = 4 - 2\cos\left(\frac{2\pi}{n}h\right) - 2\cos\left(\frac{2\pi}{n}k\right) \qquad (h, k) \in \{0, \ldots, n-1\}^2$$

with corresponding eigenvectors $x_{(v,w)}^{(h,k)} = \exp\left[i\frac{2\pi}{n}(vh+wk)\right]$. Therefore, the Green matrix can be represented as

$$(Z_C)_{(v_1,w_1)(v_2,w_2)} = \sum_{(h,k)\neq(0,0)} \frac{\exp\left[i\frac{2\pi}{n}((v_1-v_2)h+(w_1-w_2)k)\right]}{4-2\cos\left(\frac{2\pi}{n}h\right)-2\cos\left(\frac{2\pi}{n}k\right)}$$

If we consider an input current $\iota = e_{(0,0)} - e_{(\alpha,0)}$ (thus, supported on the two nodes $(0,0)$ and $(\alpha, 0)$), we obtain that the corresponding voltage is given, up to constants, by

$$\begin{aligned} W_{(v,w)} &= (Z_C)_{(v,w)(0,0)} - (Z_C)_{(v,w)(\alpha,0)} \\ &= \sum_{(h,k)\neq(0,0)} \frac{\left[1-\exp\left(-i\frac{2\pi}{n}\alpha h\right)\right]\exp\left[i\frac{2\pi}{n}(vh+wk)\right]}{4-2\cos\left(\frac{2\pi}{n}h\right)-2\cos\left(\frac{2\pi}{n}k\right)}. \end{aligned}$$

A similar explicit (but more complex) formula can be obtained for general Abelian Cayley graphs by applying Proposition 1.18. Another example is reported below.

Example 5.4 (*Hypercube*) Consider the hypercube graph H_n having node set $V = \{0,1\}^n$, defined in Example 1.3. Eigenvalues of the coincide with the numbers $2k$ for $k \in \{0,\dots,n\}$: Eigenvalue $2k$ has multiplicity $\binom{n}{k}$ and corresponding eigenvectors

$$\phi_v^{(x)} = (-1)^{\sum_i x_i v_i}, \quad x, v \in \{0,1\}^n, \sum_i x_i = k$$

Therefore,

$$(Z_C)_{vw} = \sum_{x\in\{0,1\}^n\setminus\{(0,\dots,0)\}} \frac{(-1)^{\sum_i x_i(v_i-w_i)}}{2\sum_i x_i}$$

If we consider an input current $\iota = e_{(0,0,\dots 0)} - e_{(1,1,\dots,1)}$, we obtain that the corresponding voltage is given, up to constants, by

$$W_v = \sum_{x\in\{0,1\}^n\setminus\{(0,\dots,0)\}} \frac{(-1)^{\sum_i x_i v_i} - (-1)^{\sum_i x_i(1-v_i)}}{2\sum_i x_i} \tag{5.5}$$

Even though the Green matrix is a useful tool in constructing the theory, its explicit computation can be inconvenient. However, one key advantage of the electrical network approach is that there exist simple and powerful techniques to compute voltages without the need for an explicit computation of the Green matrix. For instance, the following result collects several useful tools that permit to simplify the computation of voltages and current flows by replacing a network by an equivalent simpler one. Preliminarily, notice that also graphs with multiple edges would be appropriate in this context: Kirchoff's and Ohm's law would remain valid and the theory developed so far would directly extend to this case.

Proposition 5.1 *Let (G, C) be an electrical network. Let $\iota \in \mathbb{R}^V$ be such that $\iota^* \mathbf{1} = 0$ and let (W, ϕ) be the corresponding voltage and current flow.*

- **Parallel law**. *Suppose e_1 and e_2 are two edges insisting on the same two vertices u and v. Consider the new electrical network $(\tilde{G}, \tilde{C})$ where $\tilde{G}$ only differs from G as the two edges e_1 and e_2 are replaced by a single edge e with conductance $\tilde{C}_{ee} = C_{e_1e_1} + C_{e_2e_2}$. Then, the voltage and the current flow in $(\tilde{G}, \tilde{C})$, corresponding to the same exogenous input currents ι, coincide with W, ϕ.*
- **Series law**. *Suppose that $v \in V$ is such that $\iota_v = 0$ and $d_v = 2$ with neighbors u_1 and u_2. Consider $(\tilde{G}, \tilde{C})$ where $\tilde{G}$ is a graph on $V \setminus \{v\}$ with same undirected edges as G but $\{u_1, v\}$ and $\{v, u_2\}$ replaced by $\{u_1, u_2\}$, and $\tilde{C}_{u_1u_2} = \tilde{C}_{u_2u_1} = (C_{u_1v}^{-1} + C_{vu_2}^{-1})^{-1}$. The voltage and current flow $\tilde{W}, \tilde{\phi}$ in $(\tilde{G}, \tilde{C})$ satisfy: $\tilde{W}_w = W_w$ for every $w \in V \setminus \{v\}$, $\tilde{\phi}_{w_1w_2} = \phi_{w_1w_2}$ for nodes in $V \setminus \{v\}$ such that $\{w_1, w_2\} \neq \{u_1, u_2\}$, while $\tilde{\phi}_{u_1,u_2} = \phi_{u_1,v} = \phi_{v,u_2}$.*
- **Glueing**. *Suppose that $W_u = W_v$. Consider the new electrical network $(\tilde{G}, \tilde{C})$ where $\tilde{G}$ is obtained from G by glueing together the two nodes u and v, while $\tilde{C} = C$ maintains the same conductances an all edges, and consider the input current $\tilde{\iota}$ defined by*

$$\tilde{\iota}_w = \iota_w \ \forall w \in V \setminus \{u, v\}, \quad \tilde{\iota}_{u+v} = \iota_u + \iota_v$$

where $u + v$ denotes the glued node in $\tilde{G}$. Then, the corresponding voltage $\tilde{W}$ and current flow $\tilde{\phi}$ on $(\tilde{G}, C)$ coincide with (W, ϕ), with the only change that $\tilde{W}_{u+v} = W_u = W_v$.

Proof Straightforward check that Kirchoff's and Ohm's laws are satisfied in the new networks. □

We will see in Sects. 5.3 and 5.4 that certain applications require to assign voltages in certain nodes. Below we show how to do it. Notice first of all that (5.3) can be rewritten as

$$L(P)W = D_{C\mathbf{1}}^{-1}\iota,$$

where $P = D_{C\mathbf{1}}^{-1}C$ is the canonical reversible stochastic matrix associated with C. Componentwise, this reads as

$$W_u - \sum_{v \in V} P_{uv} W_v = \frac{\iota_u}{\sum_w C_{uw}}$$

In particular, for each $u \in V$ such that $\iota_u = 0$, it holds that

$$W_u = \sum_{v \in V} P_{uv} W_v. \tag{5.6}$$

A function $W \in \mathbb{R}^V$ satisfying (5.6) (or equivalently (5.3)) for every u belonging to a subset $\tilde{V} \subseteq V$ is said to be *harmonic* on $\tilde{V}$. We have the following result.

Proposition 5.2 (Harmonic extension) *Let $\tilde{V} \subseteq V$ and let $\tilde{W} \in \mathbb{R}^{\tilde{V}}$. Then,*

(i) There exists exactly one $W \in \mathbb{R}^V$ harmonic on $V \setminus \tilde{V}$ and such that $W_{|\tilde{V}} = \tilde{W}$.

(ii) There exists a unique $\iota \in \mathbb{R}^V$ such that $\iota^* \mathbf{1} = 0$ and $\iota_v = 0$ for every $v \notin \tilde{V}$ such that W is the voltage generated by the input current ι.

Proof (i) Order vertices of V in such a way that those in $V \setminus \tilde{V}$ appear first and consider the corresponding block decomposition

$$P = \begin{pmatrix} Q & R \\ S & T \end{pmatrix}.$$

Consider a vector of type $W = (x, \tilde{W})^*$ and impose it satisfies (5.6) for every $u \in V \setminus \tilde{V}$. This is equivalent to require

$$Qx + R\tilde{W} = x. \tag{5.7}$$

Since the graph is strongly connected, it is immediate to check, thanks to Proposition 2.4, that Q is an asymptotically stable sub-stochastic matrix. This implies that $I - Q$ is invertible, and therefore, (5.7) is solved by

$$x = (I - Q)^{-1} R\tilde{W}. \tag{5.8}$$

Consequently, $W = ((I - Q)^{-1} R\tilde{W}, \tilde{W})^*$ is the wanted harmonic extension of $\tilde{W}$.

(ii) It is sufficient to consider $\iota = L(C)W$. □

Remark 5.1 (*Harmonic extension is convex combination of assigned voltages*) The matrix $(I - Q)^{-1}R$ that appears in (5.8) has some important properties which we now discuss. First, notice that the inverse of $I - Q$ can be represented as a series $(I - Q)^{-1} = \sum_{n=0}^{\infty} Q^k$, and this implies that $(I - Q)^{-1}_{uv} \geq 0$ for all $u, v \in V \setminus \tilde{V}$. Since also R is nonnegative it follows that $(I - Q)^{-1}R$ is also nonnegative. Moreover, notice that since P is stochastic it holds $(I - Q)\mathbf{1} + R\mathbf{1} = 0$. But this implies that $(I - Q)^{-1}R\mathbf{1} = \mathbf{1}$. In other words, each row of $(I - Q)^{-1}R$ sums to 1. In general, however, we cannot say that $(I - Q)^{-1}R$ is a stochastic matrix since it is not a square matrix.

Example 5.5 (*Binary trees*) Consider a binary tree of depth t (see Fig. 5.1) with unitary conductances and assume that the root node v_0 is at voltage 0, while the 2^t leaves are at voltage 1. For symmetry reasons, all nodes at distance s from the root node have the same voltage, and thus, we can replace, by the glueing and parallel laws in Proposition 5.1, the network with an equivalent line graph L_{t+1} with set of nodes $\{v_0, v_1, \dots v_t\}$ and conductance matrix $C_{v_s v_{s+1}} = 2^{s+1}$ for $s = 0, \dots, t - 1$. This implies that the total resistance between v_0 and v_t is given by

$$\sum_{s=0}^{t-1} \frac{1}{2^{s+1}} = 1 - 2^{-t}$$

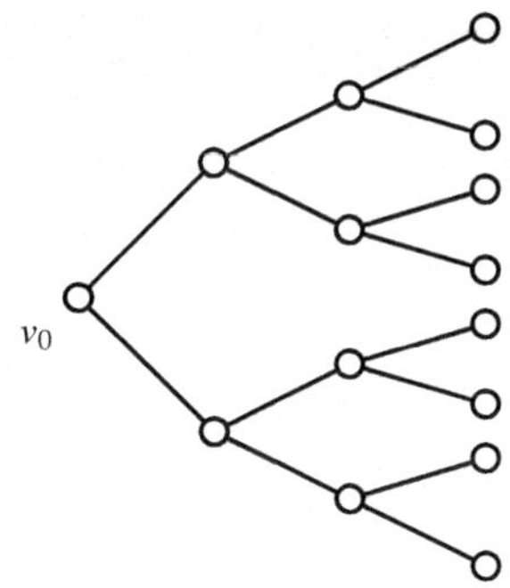

Fig. 5.1 A binary tree of depth 3: its root is labeled as v_0

The current along the line graph is thus by Ohm's law $(1-2^{-t})^{-1}$. Voltages at the various nodes can now be simply obtained by applying again Ohm's law:

$$W_{v_{s+1}} - W_{v_s} = (1-2^{-t})^{-1}2^{-(s+1)}$$

and thus,

$$W_s = (1-2^{-t})^{-1}\sum_{k=0}^{s-1} 2^{-(k+1)} = \frac{1-2^{-s}}{1-2^{-t}}$$

Example 5.6 (*Voltages in barbell graphs*) Consider two complete graphs $K_i = (V_i, E_i)$ $(i = 1, 2)$ on the set of nodes V_i and unitary conductances. Fix two nodes $v_i \in V_i$ and consider the barbell graph $G = (V, E)$ where $V = V_1 \cup V_2$ and $E = E_1 \cup E_2 \cup \{(v_1, v_2), (v_2, v_1)\}$. Consider two nodes $s_i \in (V_i \setminus v_i)$ and assign voltages $W_{s_1} = 0$ and $W_{s_2} = 1$. We want to compute the harmonic extension of W. For symmetry reasons, W will be constant at all nodes in $V_i \setminus \{s_i, v_i\}$ for $i = 1, 2$. By the glueing property and the parallel law, the electrical network can thus be replaced by a line with six nodes $s_1, w_1, v_1, v_2, w_2, s_2$ such that

$$C_{s_i,w_i} = C_{w_i,v_i} = N_i - 2\,,\ \ C_{v_1,v_2} = 1$$

where $N_i = |V_i|$. In order to compute the current, we can use the series law further reducing the electrical network to a single edge between s_1 and s_2 of conductance $[2(N_1-2)^{-1} + 2(N_2-2)^{-1} + 1]^{-1}$: The current coincides with the conductance. Using now Ohm's law we obtain

$$W_{w_1} = \frac{(N_1-2)^{-1}}{2(N_1-2)^{-1}+2(N_2-2)^{-1}+1}\,,\quad W_{v_1} = \frac{(2(N_1-2))^{-1}}{2(N_1-2)^{-1}+2(N_2-2)^{-1}+1}$$
$$W_{v_2} = \frac{2(N_1-2)^{-1}+1}{2(N_1-2)^{-1}+2(N_2-2)^{-1}+1}\,,\quad W_{w_2} = \frac{2(N_1-2)^{-1}+(N_2-2)^{-1}+1}{2(N_1-2)^{-1}+2(N_2-2)^{-1}+1}$$

5.2 Effective Resistance in Electrical Networks

A very useful concept in dealing with electrical networks is that of effective resistance between nodes. Formally, given an electrical network (G, C) and two nodes $u, v \in V$, we consider the input current $\iota = e_u - e_v$. The corresponding voltage up to translation is denoted by W, and the *effective resistance* between u and v is defined by

$$R_{\text{eff}}(u, v) := W_u - W_v.$$

The *average effective resistance* in the network is then defined as

$$R_{\text{ave}}(G, C) := \frac{1}{2N^2} \sum_{u,v \in V} R_{\text{eff}}(u, v).$$

The average effective resistance can be used as a measure of graph connectivity, in the sense that "well-connected" graphs will have small R_{ave}. We will return to this interpretation in Sect. 5.4 and in the Exercises.

Effective resistances can be characterized in terms of the Green matrix. Indeed, recalling that $W = Z_C \iota$, it holds

$$R_{\text{eff}}(u, v) = (e_u - e_v)^* Z_C (e_u - e_v) = (Z_C)_{uu} - 2(Z_C)_{uv} + (Z_C)_{vv} \tag{5.9}$$

Therefore, recalling that $Z_C \mathbf{1} = 0$, we also have

$$R_{\text{ave}}(G, C) = \frac{1}{2N^2} \sum_{u,v \in V} R_{\text{eff}}(u, v) = \frac{1}{N} \operatorname{tr}(Z_C) = \frac{1}{N} \sum_{i \geq 2} \frac{1}{\lambda_i}, \tag{5.10}$$

where $0 = \lambda_1, \ldots, \lambda_N$ are the eigenvalues of $L(C)$.

Example 5.7 (*Effective resistance on line graphs*) Consider the symmetric line graph $G = L_{N+1}$ (with vertex set $V = \{1, \ldots, N+1\}$) with conductance matrix $C \in \mathbb{R}^{V \times V}$. It immediately follows from Example 5.1 that

$$R_{\text{eff}}(1, N+1) = W_{N+1} - W_1 = \sum_{k=1}^{N} C_{kk+1}.$$

In the special case when $C = A_G$ (all edges have conductance equal to 1), we obtain $R_{\text{eff}}(1, N+1) = N$.

Example 5.8 (*Effective resistance on trees*) Let $G = (V, E)$ be a tree, C a conductance matrix and $\iota \in \mathbb{R}^V$ an input current such that $\iota_v = 1 = -\iota_w$ while $\iota_u = 0$ for every $u \neq v, w$. Consider the only path $v = v_1, \ldots, v_{N+1} = w$ connecting v to w in G. From Example 5.2 and a repetition of glueing operations, it is immediate to check that all edges not contained in this path will have a current flow equal to 0. Consequently, a current flow equal to 1 will be flowing from v to w along the connecting path as if it was a line graph. From Example 5.7, it thus follows that $R_{\text{eff}}(v, w) = \sum_{k=1}^{N} C_{v_k v_{k+1}}$ In the special case when $C = A_G$, it follows that the effective resistance between two nodes coincides with their distance on the tree. Notice that it is also easy to compute the corresponding voltage at any vertex of the tree. Given a vertex $u \in V$, let v_k be the closest vertex of the path $v = v_1, \ldots, v_{N+1} = w$ to u. Then, $W_u = W_{v_k}$.

Example 5.9 (*Effective resistance on cycles*) Consider the graph C_n with node set $\mathbb{Z}_n$. By applying the parallel law and Example 5.7, we observe

$$R_{\text{eff}}(u, v) = \left(|v - u|^{-1} + (n - |v - u|)^{-1}\right)^{-1} = \frac{|v - u|\,(n - |v - u|)}{n}.$$

For general graphs, the computation of the effective resistance can be a complex problem and closed formulas can hardly be found. However, there are tools to efficiently estimate it. Before we can illustrate them, we need to introduce a further concept. Given an electrical circuit $(G = (V, E), C)$, a *flow* on it is any function $\phi \in \mathbb{R}^E$ such that $\phi_{uv} = -\phi_{vu}$. As before $\bar{E}$ will denote the set of undirected edges of G. Given an incidence matrix B of G, we can consider the flow defined on $\bar{E}$ and denote it by $\bar{\phi}$ (as we did for the current flow above). The *energy* of a flow ϕ is defined as

$$|\phi| = (1/2) \sum_{(uv) \in E} \frac{\phi_{uv}^2}{C_{uv}} = \sum_{\{uv\} \in \bar{E}} \frac{\bar{\phi}_{\{uv\}}^2}{C_{uv}}$$

Given $\iota \in \mathbb{R}^V$ such that $\iota^* \mathbf{1} = 0$, we say that a flow ϕ is compatible with ι if it satisfies Kirchoff's law $B^* \bar{\phi} = \iota$. The following variational principle holds true (a proof can be found in [20, Theorem 9.10]):

Lemma 5.1 (Thomson's principle) *Let* (G, C) *be an electrical network. Then,*

$$R_{\text{eff}}(u, v) = \inf\{|\phi| \;:\; \phi \text{ is a flow compatible with } \iota = e_u - e_v\}.$$

Moreover the unique minimizer is the current flow induced by the input current $\iota = e_u - e_v$.

An immediate important consequence is the following result.

Corollary 5.1 (Raileigh's monotonicity law) *Let* G *be a symmetric graph and* C' *and* C'' *two conductance matrices on* G *such that* $C'_{uv} \leq C''_{uv}$ *for all* $u, v \in V$. *Then, for any pair of vertices the corresponding effective resistances in the two networks satisfy*

$$R'_{\text{eff}}(u, v) \geq R''_{\text{eff}}(u, v).$$

Glueing nodes in an electrical network is equivalent to put conductance equal to ∞ between certain pairs of nodes. By virtue of Raileigh monotonicity law, this implies that the effective resistance, in the glueing operation, can never increase. The following is an example of application of this useful remark.

Example 5.10 (*Effective resistance on grids*) Consider a bidimensional grid $L_n \times L_n$ with set of nodes $\{1, \ldots, n\}^2$ and unitary conductances. Suppose we want to estimate the effective resistance between $(1, 1)$ and (n, n). Replace the network by a line network obtained by glueing together all nodes at distance d from $(1, 1)$ (and denote such super node by v_d). The nodes $(1, 1)$ and (n, n) become v_0 and v_{2n-2} in the new network. Let n_d be the number of nodes at distance d from $(0, 0)$. Since (x, y) is at distance d from $(0, 0)$ if and only if $x + y = d + 2$, we have that $n_d = d + 1$ if $d \leq n - 1$. It follows that $C_{v_d v_{d+1}} = 2(d + 1)$ for all $d = 0, \ldots, n - 2$. Considering that the new network is specularly symmetric with respect to the node v_{n-1}, we have that

$$R_{\text{eff}}((1, 1), (n, n)) \geq R_{\text{eff}}(v_0, v_d) = \sum_{d=0}^{n-2} \frac{1}{d+1} \geq \int_1^n \frac{1}{x} \mathrm{d}x = \log n.$$

By constructing suitable flows and applying Thompson's principle, it can be shown that that it also holds $R_{\text{eff}}((1, 1), (n, n)) \leq 2 \log n$. In case of grids of higher dimension L_n^d, instead, there exists $c_d > 0$ such that $R_{\text{eff}}(v, w) \leq c_d$ for all $v, w \in \{1, \ldots, n\}^d$. More details can be found in [20].

When voltages are imposed at some nodes, all other voltages can be computed in terms of effective resistances.

Proposition 5.3 (Voltages and effective resistance) *Let (G, C) be an electrical network and v_0 and v_1 two distinct nodes in G. Let W be the voltage satisfying $W_{v_0} = 0$ and $W_{v_1} = 1$. Then,*

$$W_v = \frac{1}{2} + \frac{R_{\text{eff}}(v, v_0) - R_{\text{eff}}(v, v_1)}{2R_{\text{eff}}(v_0, v_1)} \quad \forall v \in V. \tag{5.11}$$

Proof Clearly, we can represent $W = Z_C\iota + c\mathbf{1}$ where ι is the input current given by $\iota = R_{\text{eff}}(v_0, v_1)^{-1}[e_{v_1} - e_{v_0}]$ and c a constant. Imposing $W_{v_0} = 0$, we obtain that $c = R_{\text{eff}}(v_0, v_1)^{-1}[(Z_C)_{v_0 v_0} - (Z_C)_{v_0 v_1}]$. For a generic $v \in V$, the voltage can thus be computed as follows (denoting Z_C as Z for conciseness):

$$\begin{aligned} W_v =& \frac{Z_{vv_1} - Z_{vv_0} + Z_{v_0v_0} - Z_{v_0v_1}}{R_{\text{eff}}(v_0, v_1)} \\ =& \frac{(-Z_{vv} + 2Z_{vv_1} - Z_{v_1v_1}) + (Z_{vv} - 2Z_{vv_0} + Z_{v_0v_0}) + (Z_{v_0v_0} - 2Z_{v_0v_1} + Z_{v_1v_1})}{2R_{\text{eff}}(v_0, v_1)}. \end{aligned}$$

The result now follows from relation (5.9). □

We now present a significant example.

Example 5.11 (*Harmonic functions on line graphs*) Consider the symmetric line graph $G = L_{n+1}$ with vertex set $V = \{1, \ldots, n + 1\}$ and with conductance matrix

$C = A_G$. Let $\tilde{V} = \{1, n+1\}$ and put $\tilde{W}_1 = 0$ and $\tilde{W}_{n+1} = 1$. Let W be the harmonic extension of $\tilde{W}$. Using formula (5.11), we immediately obtain

$$W_k = \frac{1}{2}\frac{(k-1) - (n+1-k) + n}{n} = \frac{k-1}{n}, \ \forall k.$$

Notice that if the voltage assigned in nodes 1 and $n+1$ were different, namely $\tilde{W}_1 = \tilde{w}_1$ and $\tilde{W}_{n+1} = \tilde{w}_{n+1}$, then using the fact that the harmonic extension is a linear function of the boundary conditions, we would obtain the new voltage

$$\tilde{W}_k = \tilde{w}_1 + \frac{k-1}{n}(\tilde{w}_{n+1} - \tilde{w}_1).$$

Furthermore, it is important to be aware that Proposition 5.3 can be applied when the two nodes v_0 and v_1 are the outcome of glueing operations. Hence, its scope of application covers all cases where some nodes are connected to any *two* voltage levels.

5.3 Averaging Dynamics with Stubborn Agents

In the examples of consensus models studied so far, we have essentially assumed that all the agents are implementing the same dynamic law, all of them cooperating to reach a consensus. Very interesting models can however be obtained considering instead heterogeneous model where agents have different behaviors. As a special case, here we investigate the case when some of the agents maintain fixed initial state. These agents will be called *stubborn*. Several interpretations are possible. In robotic networks, these agents can be interpreted as leaders who are trying to keep the rest of the units within a certain region: In this context, we talk about the *containment* problem. In the context of opinion dynamics, stubborn agents play the role of opinion leaders or influencers.

Let $G = (V, E)$ be a strongly connected aperiodic graph endowed with a stochastic matrix $P \in \mathbb{R}^{V\times V}$ such that $G_P = G$. Consider the split $V = V^{\ell} \cup V^{f}$ with the understanding that agents in V^{ℓ} are the *leaders* while those in V^{f} are the *followers*. The dynamics we want to consider is given by the modified stochastic matrix $\tilde{P} \in \mathbb{R}^{V\times V}$ defined by

$$\tilde{P}_{uv} = \begin{cases} P_{uv} \text{ if } u \in V^f,\ v \in V \\ \delta_{uv} \text{ if } u \in V^{\ell} \end{cases}$$

If $V^{\ell} = \{v^{\star}\}$, the node $v^{\star}$ will be globally reachable and aperiodic for the graph $G_{\tilde{P}}$. Therefore, thanks to Theorem 2.2, there will be convergence to a consensus. Because of Proposition 2.5, it follows that the corresponding invariant probability for P will be $\pi = \delta_{v^{\star}}$ and therefore

$$\tilde{P}^t x(0) \to \mathbf{1} x(0)_{v^\star}$$

In other terms, consensus coincides with the initial (unchanged) state of the unique leader $v^\star$. If $|V^\ell| > 1$, the graph $G_{\tilde{P}}$ will not possess a globally reachable vertex, and therefore, consensus will not in general be achievable. Nevertheless, we would like to understand the behavior of $\tilde{P}^t x(0)$ for $t \to +\infty$.

If we order elements in V in such a way that followers come first, the matrix $\tilde{P}$ will have the block structure:

$$\tilde{P} = \begin{bmatrix} Q & R \\ 0 & I \end{bmatrix}$$

where $Q \in \mathbb{R}^{V^f \times V^f}$, $R \in \mathbb{R}^{V^f \times V^\ell}$, and where $I \in \mathbb{R}^{V^\ell \times V^\ell}$ is the identity matrix. If we split accordingly the state vector $x(t) = (x^f(t), x^\ell(t) \in \mathbb{R}^V$, we thus have dynamics

$$\begin{aligned} x^f(t+1) &= Qx^f(t) + Rx^\ell(t) \\ x^\ell(t+1) &= x^\ell(t) \end{aligned} \tag{5.12}$$

By the assumption made, it follows that Q is sub-stochastic satisfying the assumptions of Proposition 2.4. Hence, Q is asymptotically stable. These dynamics easily imply that $x^f(t)$ converges to $x^f(\infty) \in \mathbb{R}^{V^f}$ determined by the fixed point relation:

$$x^f(\infty) = Qx^f(\infty) + Rx^\ell(0)$$

which is equivalent to

$$(I - Q)x^f(\infty) = Rx^\ell$$

or since $I - Q$ is invertible, to

$$x^f(\infty) = (I - Q)^{-1} R x^\ell \tag{5.13}$$

where $x^\ell = x^\ell(0)$. Notice in particular that the initial condition of the state of the followers, $x^f(0)$ does not play any role in the final state. If all the leaders share the same state, $x^\ell_v = c$ for every $v \in V^\ell$, then it is immediate to check that $x^f(\infty)_v = c$ for every $v \in V^f$, namely, they reach consensus. In general, however, $x_f(\infty)$ is not a consensus state. If we confront the formula for $x^f(\infty)$ above with (5.8), we see that, indeed, $x^f(\infty)$ can be interpreted as the harmonic extension of the leader assignment x^ℓ. If the original matrix P is a reversible stochastic matrix, we can then apply all the machinery from electrical networks for computing the vector $x^f(\infty)$.

Example 5.12 Consider the graph L_{N+1} with vertex set $\{1, \dots, N+1\}$ and leader nodes $\ell_1 = 1$ and $\ell_2 = N + 1$. Let P be the SRW on the follower nodes. It follows from Example 5.11 that the followers' limit state is given by

$$x_k^f(\infty) = \frac{x_{\ell_2}^\ell - x_{\ell_1}^\ell}{N}(k-1) + x_{\ell_1}^\ell.$$

This formula also shows that there is consensus if and only if $x_{\ell_1}^\ell = x_{\ell_2}^\ell$.

Remark 5.2 (*Connectivity and influence*) Notice that $\left((I-Q)^{-1}R\right)_{hk} = \sum_n \left(Q^n R\right)_{hk}$ is not equal to 0 if and only if there exists a path in the graph connecting the follower h to the leader k. This implies that if a follower h can reach leader k only, then $x_h^f(\infty) = x_k^\ell(0)$.

Remark 5.3 (*Multi-dimensional case*) If the evolving state of each unit $x_v(t)$ is a vector (e.g., in $\mathbb{R}^2$ or $\mathbb{R}^3$) possibly indicating positions, the considerations above remain valid with the usual interpretation of the matrix multiplications done in previous chapter. Relation (5.13), in this case, has an even more vivid geometric representation. It says that the asymptotic state of each follower is a convex combination of the state of the leaders, in other words, each follower will eventually stay in the convex polyhedron whose vertices are the states of the leaders.

5.4 Estimation from Relative Measurements

Consider a set of agents V and a symmetric connected graph $G = (V, E)$. Each agent v possesses an attribute $\bar{x}_v \in \mathbb{R}^q$ (to be interpreted as position or quality, for instance) which is unknown to the agent itself. Any pair of agents $v, w \in V$, connected by an edge in G, make a cooperative measurements of their relative position

$$b_{\{v,w\}} = \bar{x}_v - \bar{x}_w + n_{\{v,w\}}, \tag{5.14}$$

where $n_{\{v,w\}}$ is a random variable modeling the measurement noise. We will assume that random variables are independent and identically distributed with mean 0 and variance σ^2. Also, we will assume that $q = 1$: This choice does not entail any loss of generality as the case of $q > 1$ can be captured working componentwise.

Notice that the measurements and the noises are naturally defined on the set of undirected edges $\bar{E}$ of the graph $b, n \in \mathbb{R}^{\bar{E}}$. However, also notice that the measurement model (5.14) assumes that a direction has been decided at the level of each pair v, w. If we consider the incidence matrix B of G corresponding to such chosen directions (e.g., $B_{\{v,w\}v} = 1$ in reference to (5.14)), then we can rewrite relations (5.14) in a more compact form as

$$b = B\bar{x} + n. \tag{5.15}$$

On the basis of the available measurements b, the goal is to obtain an estimate $\hat{x}$ of $\bar{x}$. A classical solution is the so called *least squares estimator*, defined as

$$\hat{x} := \operatorname{argmin}_{x \in \mathbb{R}^V} ||Bx - b||_2^2 \tag{5.16}$$

Notice that for sure the above minimum is not unique as $B\mathbf{1} = 0$. Indeed, notice that any translation of all real positions $\bar{x}$ of a vector $c\mathbf{1}$ would not change the vector b: In other terms, estimation can only be achieved modulo a translation addend. We will see below that this is the only "freedom" in the system so that (5.16) is uniquely defined up to this translation.

Remark 5.4 (*Maximum likelihood interpretation*) In the case when the variables $n_{\{v,w\}}$ are Gaussian, the least squares estimator coincide with the classical Maximum Likelihood (ML) estimator. Indeed, the density function of $b = Bx + n$ assuming x to be a parameter is given by

$$f(b|x) = \frac{1}{(2\pi\sigma^2)^{|\bar{E}|/2}} e^{-\frac{||b-Bx||_2^2}{(2\sigma^2)|\bar{E}|}}$$

and therefore, the ML estimator is given by

$$\hat{x}_{\text{ML}} := \operatorname{argmax}_{x \in \mathbb{R}^V} f(b|x) = \operatorname{argmin}_{x \in \mathbb{R}^V} ||Bx - b||_2^2$$

Consider the functional

$$J(x) = ||Bx - b||_2^2, \tag{5.17}$$

which is what we want to minimize. Notice that $J(x) = x^* B^* Bx - 2b^* Bx + ||b||_2^2$ is indeed a convex quadratic function and its minima coincide with its stationary points. Since its gradient is given by $\nabla J(x) = 2B^* Bx - 2B^* x$, its minima are the solutions of the equation $L_G x = B^* b$ (recall that $B^* B = L_G$). This equation is the voltage equation in the electrical network (G, A_G) and with input current $\iota = B^* b$ (notice that $\iota^* \mathbf{1} = b^* B\mathbf{1} = 0$ as required). Solutions are then given by

$$\hat{x} = Z_G B^* b + c\mathbf{1}, \tag{5.18}$$

provided we denote $Z_G = Z_{A_G}$. Notice that, since $\mathbf{1}^* Z_G = 0$, it follows that the solution $\hat{x} = Z_G B^* b$ is the (only) one satisfying $\mathbf{1}^* \hat{x} = 0$ (barycenter in the origin).

Remark 5.5 (*Trees*) In the special case when G is a tree, notice that B is an $(N - 1) \times N$-matrix having rank equal to $N - 1$. Therefore, B is onto and there must exist $\hat{x}$ satisfying $B\hat{x} = b$; this is for sure a minimizer of $J(x)$, hence it must coincide with the solution (5.18). This implies that must hold $BZ_G B^* = I$.

We now want to study the performance of the least squares estimator. Particularly, we are interested in evaluating the effects of the noise and the topology of the graph. A natural performance measure is the minimum mean quadratic error, which considering the nonuniqueness of the solution, takes the form

$$J_{\text{rel}} = \frac{1}{N} \mathbb{E} \|\hat{x} - \bar{x}\|_2^2 := \frac{1}{N} \min_{c \in \mathbb{R}} \mathbb{E} ||(Z_G B^* b + c\mathbf{1}) - \bar{x}\|_2^2, \tag{5.19}$$

where the expectation is taken over the noise n. Notice that the optimal c is given by $c = N^{-1}\mathbf{1}^*\bar{x}$, which corresponds to the estimate $\hat{x}$ having the same barycenter of x.

The following simple result shows that the mean-square error of the least squares solution depends only on the variance of the noise and on the topology of the graph through the Green matrix of the graph:

Proposition 5.4 (MSE formula) *Provided the undirected graph G is connected, cost* (5.19) *can be computed as*

$$J_{\text{rel}} = \frac{\sigma^2}{N}\operatorname{tr}(Z_G). \tag{5.20}$$

Proof We compute as follows

$$\begin{aligned} J_{\text{rel}} &= \frac{1}{N}\min_{c\in\mathbb{R}} \mathbb{E}\|(Z_G B^* b + c\mathbf{1}) - \bar{x}\|_2^2 \\ &= \frac{1}{N}\min_{c\in\mathbb{R}} \mathbb{E}\|Z_G L_G \bar{x} - \bar{x} + c\mathbf{1} + Z_G B^* n\|_2^2 \\ &= \frac{1}{N}\min_{c\in\mathbb{R}} \mathbb{E}\|\mathbf{1}(-N^{-1}\mathbf{1}^*\bar{x} + c) + Z_G B^* n\|_2^2 \\ &= \frac{1}{N}\min_{c\in\mathbb{R}} \mathbb{E}\left[N(-N^{-1}\mathbf{1}^*\bar{x} + c) + n^* B Z_G^2 B^* n\right] \\ &= \frac{1}{N}\mathbb{E}[n^* B Z_G^2 B^* n] = \frac{1}{N}\mathbb{E}\operatorname{tr}[Z_G B^* n n^* B Z_G] = \frac{1}{N}\operatorname{tr}[Z_G B^* \mathbb{E}[n n^*] B Z_G] = \frac{\sigma^2}{N}\operatorname{tr}(Z_G), \end{aligned}$$

by using (5.4) and recalling $N^{-1}\mathbf{1}^*\bar{x} = c$. □

It follows from (5.10) that

$$J_{\text{rel}} = \sigma^2 R_{\text{ave}}(G, A_G).$$

Whenever we are able to compute or estimate the effective resistance in a graph, we will be able to estimate the performance of the mean-square position estimator. In particular, it follows from Examples 5.7 and 5.10 that for d-dimensional grids

$$J_{\text{rel}} = \begin{cases} \Theta(N) & \text{for } d = 1 \\ \Theta(\log N) & \text{for } d = 2 \\ \Theta(1) & \text{for } d > 2 \end{cases} \tag{5.21}$$

as $N \to \infty$. We thus have strikingly different behaviors of the algorithm for $d \le 2$ and $d > 2$, as in the first case performance degrades as N increases. A similarly poor performance affects trees: Example 5.8 implies that J_{rel} is linear in the diameter on trees with bounded degrees. This fact means that, even though the tree structure is sufficient to estimate absolute distances (as explained in Remark 5.5), the availability of additional measurements is essential to obtain good performance.

As shown by (5.18), the solution to the optimization problem (5.16) is easily obtained analytically. However, it is not immediately clear whether in practice such a solution can be computed in a distributed fashion by the nodes. We show now that the answer is positive, by presenting a distributed algorithm that allows each node v to compute its own component of the estimate $\hat{x}_v$.

Let us recall that $J(x)$ as defined in (5.17) is a convex function: Then its minima can be found by an iterative gradient descent algorithm. Let $x(t) \in \mathbb{R}^V$ be the vector of node estimates at iteration t. Then, we consider the following algorithm

$$x(t+1) = x(t) - \tau \nabla J(x(t)),$$

with $\tau > 0$ to be determined in order to ensure convergence. The recursive law can be rewritten as:

$$\begin{aligned} x(t+1) &= x(t) - \tau(L_G x(t) - B^* b) \\ &= (I - \tau L_G)x(t) + \tau B^* b \end{aligned}$$

Defining $P := I - \tau L_G \in \mathbb{R}^{V \times V}$ and $y := \tau B^* b \in \mathbb{R}^I$, we obtain the compact form

$$x(t+1) = Px(t) + y. \qquad (5.22)$$

It is of note that the matrix P is inherently adapted to the measurement graph G, in the sense that $P_{uv} > 0$ only if (u, v) is an edge in G. This observation is key as it implies that the algorithm is naturally distributed over the graph which describes the problem, that is, there is *no need for communication between agents which do not share a measurement.*

The convergence properties of the algorithm are summarized in the next result.

Proposition 5.5 (Convergence) *Let G be symmetric and strongly connected. Choose* τ *such that* $0 < \tau < \frac{1}{d_{\max}}$*, where* $d_{\max}$ *denotes the largest degree in G. Then, the algorithm* (5.22) *is such that*

$$\lim_{t \to +\infty} x(t) = \hat{x},$$

where $\hat{x}$ *is the solution in (5.18) characterized by the condition* $\frac{1}{N}\mathbf{1}^*\hat{x} = \frac{1}{N}\mathbf{1}^* x(0)$.

Proof From the assumption on τ it follows that P is a symmetric, irreducible, aperiodic stochastic matrix. Then, we know from Chap. 2 that 1 is a simple eigenvalue whose eigenspace is spanned by $\mathbf{1}$, while all other eigenvalues are, in modulus, strictly less than 1. Since $\mathbf{1}^* y = 0$, it easily follows that $x(t)$ converges to a solution of the equation $x = Px + y$. Substituting the expression of P, we immediately get the result. Invariance of the barycenter simply follows by applying $\mathbf{1}^*$ to both sides of (5.22). □

We observe that, given an initial condition $x(0)$, the algorithm converges to a corresponding solution $\hat{x}$, specifically that one with the same average as $x(0)$. Then, in order to converge to the best solution, it is necessary to impose the same average of $\bar{x}$ to the initial condition $x(0)$.

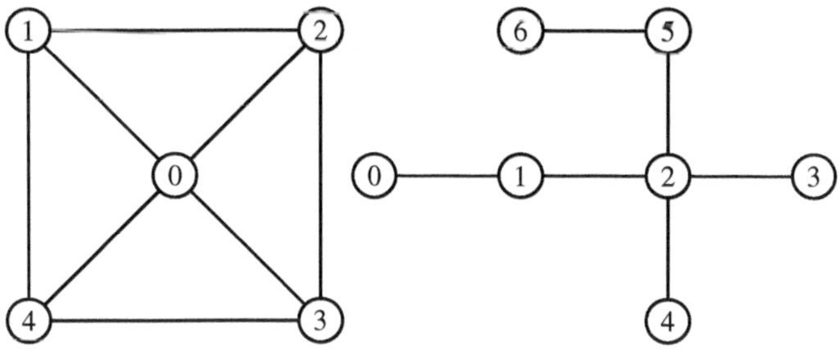

Fig. 5.2 Graphs G_1 and G_2 for Exercise 5.2

Exercises

Electrical Networks

Exercise 5.1 (*Notions of incidence matrix*) Compare the definition of incidence matrix given at the beginning of this chapter for undirected graphs and the notion defined in Exercise 1.12 for general weighted graphs.

Exercise 5.2 (*Potentials on small graphs*) Consider the graphs is Fig. 5.2. Compute the voltages

(i) W' defined on G_1 such that $W'_0 = 0$ and $W'_1 = 1$;
(ii) W'' defined on G_1 such that $W''_0 = 3$ and $W''_1 = 1$;
(iii) W''' defined on G_2 such that $W'''_0 = 0$ and $W'''_3 = 1$ and $W'''_6 = 2$.

Exercise 5.3 (*Voltages on a hypercube*) Consider the hypercube graph H_n having node set $V = \{0, 1\}^n$, defined in Example 1.3, with unitary conductances. Consider an input current $\iota = e_{(0,0,\dots,0)} - e_{(1,1,\dots,1)}$. Voltages at various nodes can be computed using the following method which is alternative to Example 5.4. First notice that for symmetry reasons, nodes at a certain distance d from $(0, 0, \dots 0)$ will all have the same voltage. Transform consequently the electrical network into an equivalent line and compute voltages. Show that you obtain the same result as formula (5.5).

Exercise 5.4 (*Effective resistance is a distance*) Verify that the effective resistance satisfies the axioms (recalled in Exercise 1.11) to be a metric on the set of nodes of an electrical network.

Exercise 5.5 (*Average effective resistance*) Using either (5.20) or an "electrical" argument, compute $R_{\text{ave}}(G)$ of the following graphs (assume for simplicity to have unitary conductances):

(i) complete graph K_N;
(ii) cycle graph C_N;
(iii) complete bipartite graph K_{N_1,N_2};

(iv) hypercube graph H_k;
(v) barbell graph as defined in Example 5.6;
(vi) (toroidal) grid of dimension d, thereby proving (5.21).

Exercise 5.6 (*Effective resistance on binary tree*) Consider a binary tree of depth n with edges of unit conductance. Compute the effective resistance between the root and the leaves (glued together).

Exercise 5.7 (*Foster's equality*) Let the conductance matrix C have unit entries. Show that $\sum_{\{u,v\}\in\bar{E}} R_{\text{eff}}(u,v) = |V| - 1$.

Exercise 5.8 (*Green matrix of a stochastic matrix*) For any aperiodic irreducible stochastic matrix P having invariant probability π, the Green matrix can be defined as

$$Z_P := \sum_{t=0}^{+\infty} (P^t - \mathbf{1}\pi^*).$$

Show that $Z_P L(P) = L(P) Z_P = I - \mathbf{1}\pi^*$ and $(Z_P + \mathbf{1}\pi^*) = (L(P) + \mathbf{1}\pi^*)^{-1}$.

Exercise 5.9 (*J_x cost on symmetric matrices*) Let the N-dimensional stochastic irreducible aperiodic matrix P be symmetric and recall the cost

$$J_x(P) = \frac{1}{N} \operatorname{tr} \sum_{t\geq 0} (P^{2t} - \frac{1}{N}\mathbf{1}\mathbf{1}^*)$$

defined in (4.7). Verify that, following the notation from Exercise 5.8,

$$J_x(P) = \frac{1}{N} \operatorname{tr} Z_{P^2} = \frac{1}{N} \sum_{i=2}^{N} \left(1 - \mu_i(P^2)\right)^{-1},$$

and thus by virtue of (5.10)

$$J_x(P) = R_{\text{ave}}(\mathcal{G}_{P^2}, P^2).$$

Exercise 5.10 (*J_x cost on reversible matrices*) This exercise extends Exercise 5.9 to *reversible* irreducible aperiodic matrices. In that case, the cost takes the form $J_x(P) = \frac{1}{N} \sum_{t\geq 0} ||P^t - \mathbf{1}\pi^*||_F^2$. For a reversible P, we can define the associated conductance matrix as

$$\Phi(P) = N \operatorname{diag}(\pi) P.$$

(i) Verify that $\Phi(P)$ is symmetric and $\mathbf{1}^*\Phi(P)\mathbf{1} = N$.
(ii) Verify that if C is a conductance matrix, then $\Phi(D_{C\mathbf{1}}^{-1} C) = (\mathbf{1}^* C \mathbf{1})^{-1} N\, C$.
(iii) [22, Theorems 3.1 and 3.2] If we assume that P is reversible and irreducible with positive diagonal, we let $D = \Phi(P^2)$ and we denote by A the adjacency matrix of $\mathcal{G}_P$, then

$$\frac{N^2\pi_{\min}^3}{\pi_{\max}} R_{\text{ave}}(\mathcal{G}_D, D) \leq J_x(P) \leq \frac{N^2\pi_{\max}^3}{\pi_{\min}} R_{\text{ave}}(\mathcal{G}_D, D),$$

and

$$\frac{N\pi_{\min}^3}{8p_{\max}^2 d_{\max}^2 \pi_{\max}^2} R_{\text{ave}}(\mathcal{G}_P, A) \leq J_x(P) \leq \frac{N\pi_{\max}^3}{8p_{\min}^2 \pi_{\min}^2} R_{\text{ave}}(\mathcal{G}_P, A),$$

where $\pi_{\min} \leq \pi_v \leq \pi_{\max}$ and $d_v \leq d_{\max}$ for every node v and $p_{\min} \leq P_{uv} \leq p_{\max}$ for every $P_{uv} > 0$.

Exercise 5.11 (*J_x cost on example graphs*) Derive the scaling of J_x for the symmetric random walk matrix on the graphs of Exercise 5.5. To this goal, you can apply Exercises 5.9 or 5.10 depending on the graph.

Consensus with stubborn agents

Exercise 5.12 (*Asymptotic followers state*) Consider the two graphs (with leader and follower nodes) in Fig. 5.3, and the matrices corresponding to the Simple Random Walks on them.

(i) Referring to the graph in Fig. 5.3 (left), compute the limit states of the followers as a function of the states of the leaders l_0, l_1, l_2, l_3, l_4, as time goes to infinity.
(ii) Referring to the graph in Fig. 5.3 (right), compute the limit states of the followers as a function of the states of the leaders l_0 and $l_1 = l_2 = l_3 = l_4$. (Hint: take advantage of the symmetries to reduce the number of unknowns).

Exercise 5.13 (*Asymptotic followers state on trees*) Compute $x^f(\infty)$ as a function of $x^\ell(0)$ for the SRW on the trees with stubborn nodes in Fig. 5.4 Generalize the second graph when the lines departing from the common central node have any length n.

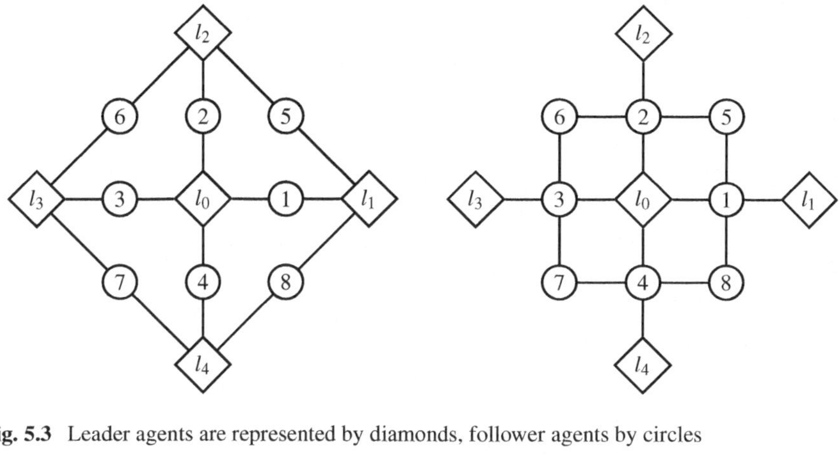

Fig. 5.3 Leader agents are represented by diamonds, follower agents by circles

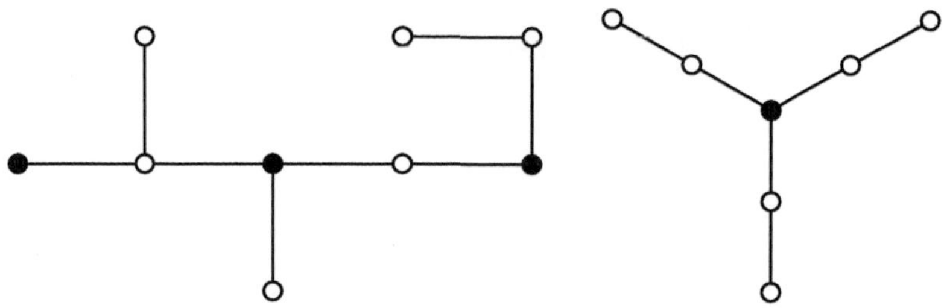

Fig. 5.4 Trees with stubborn agents (filled in black)

Exercise 5.14 (*Leaders in barbell graphs I*) Consider the graph $G = (V, E)$ where $V = V' \cup V''$ with $V' = \{v'_1, \ldots, v'_N\}$ and $V'' = \{v''_1, \ldots, v''_M\}$ and

$$E := \{(v'_i, v'_k) \mid i, k = 1, \ldots, N\} \cup \{(v''_i, v''_k) \mid i, k = 1, \ldots, M\} \cup \{(v'_1, v''_1), (v''_1, v'_1)\}$$

Assume that v'_a and v''_a are two leaders having opinion, respectively, equal to 0 and 1. Assume for the remaining nodes the consensus dynamics induced by the natural SRW and consider the asymptotic opinions as time goes to infinity.

(i) Compute the asymptotic opinions of the followers.
(ii) Compute the limits of such asymptotic opinions in the case when $N = M \to +\infty$.
(iii) Compute the limits of such asymptotic opinions in the case when $N = M^2 \to +\infty$.

Exercise 5.15 (*Leaders in barbell graphs II*) Consider the graph $G = (V, E)$ where $V = V' \cup V''$ with $V' = \{v'_1, \ldots, v'_N\}$ and $V'' = \{v''_1, \ldots, v''_N\}$ and

$$E := \{(v'_i, v'_k) \mid i, k = 1, \ldots, N\} \cup \{(v''_i, v''_k) \mid i, k = 1, \ldots, M\} \\ \cup \{(v'_h, v''_h), (v''_h, v'_h) \mid h = 1, \ldots r\}$$

Assume that $\{v'_{r+1}, \ldots v'_{r+s}\}$ and $\{v''_{r+1}, \ldots v''_{r+s}\}$ are two set of leaders having opinion, respectively, equal to 0 and 1. Consider for the remaining nodes the consensus dynamics induced by a SRW on this graph.

(i) Compute the asymptotic opinions of the followers as functions of r and s.
(ii) Compute the limits of such asymptotic opinions in the case when $N \to +\infty$ and r, s are kept constant.
(iii) Compute the limits of such asymptotic opinions in the case when $\lceil r = \alpha N \rceil$, $\lceil s = \beta N \rceil$, and $N + \infty$.

Exercise 5.16 (*Optimal leader selection*) Let x denote the harmonic extension on a graph when $V^\ell = \{v_0, v_1\}$ and $x_{v_i} = i$. We think of the two leaders as competing to maximize their own influence. We define

$$H(v_0, v_1) = \frac{1}{N} \sum_{v \in V} x_v.$$

Note that $H(v_0, v_1) \in (0, 1)$.

(i) Show that

$$H(v_0, v_1) = \frac{1}{2} - \frac{\frac{1}{N}\sum_v R_{\text{eff}}(v, v_0) - \frac{1}{N}\sum_v R_{\text{eff}}(v, v_1)}{2R_{\text{eff}}(v_0, v_1)}$$

(ii) Consider the optimization problem

$$\min_{v_0 \in V} \max_{v_1 \in V \setminus \{v_0\}} H(v_0, v_1).$$

Show that the optimal value is not larger than $\frac{1}{2}$ and the optimal solution is the optimal solution of $\min_w \sum_v R_{\text{eff}}(v, w)$.

Bibliographical Notes

The most influential works for our treatment of electrical networks in connection with reversible stochastic matrices are the classical monograph [11] and the textbook [20], where the reader can find a more comprehensive treatment. A detailed analysis of the average resistance on d-dimensional graphs, which refines and extends Example 5.10 and Eq. (5.21), has been provided in various papers [2, 8, 31, 33]: For instance, it is known that the average resistance decreases with increasing d. The optimal choice of how to distribute conductances to minimize the average resistance is studied in [17].

Consensus with stubborn agents has attracted significant attention, in view of different applications. In robotic networks, it can be seen as a containment problem [19]. In social networks, stubborn agents that do not change their opinions are present, explicitly or implicitly, in a variety of models of opinion dynamics [1, 14, 16, 23, 25]. A recent survey of related literature has been given in [26]. Electrical networks have been used as a tool in this context by [10, 34]. Recently, the problem of the optimal placement of stubborn agents has recently attracted significant attention, also in relation with classical problems of actuator selection in control theory: Various objective functions have been considered, see [9, 12, 21]. The formulation used in Exercise 5.16 derives from [34, 35]. The paper [34] proposes a message-passing algorithm to effectively solve the placement problem for v_1 (after v_0 is in place). The algorithm was deduced within the electrical framework but has now been extended to nonreversible update matrices [29].

The problem of estimation from relative measurements studied in Sect. 5.4 has been brought to our attention by reading [3–5]. The electrical framework is a useful tool for its analysis [4, 32]. This estimation problem relates to various streams of applied research: It can be interpreted as a problem of relative localization between mobile robots [3], sensor calibration for wireless sensor networks [6], statistical ranking in machine learning [24], clock synchronization [18], or voltage estimation in power networks [13].

The simple distributed gradient algorithm is analyzed in [30], but several more sophisticated solutions have been proposed since at least [3]. We note that also randomized dynamics have been studied, which extend the ideas of Chap. 3 to consensus-like dynamics with inputs [7, 15, 27]. A general convergence analysis of such affine consensus-like randomized dynamics is given in [28].

References

1. Acemoglu, D., Como, G., Fagnani, F., Ozdaglar, A.: Opinion fluctuations and disagreement in social networks. Math. Oper. Res. **38**(1), 1–27 (2013)
2. Bamieh, B., Jovanovic, M.R., Mitra, P., Patterson, S.: Coherence in large-scale networks: dimension-dependent limitations of local feedback. IEEE Trans. Autom. Control **57**(9), 2235–2249 (2012)
3. Barooah, P., Hespanha, J.P.: Estimation from relative measurements: algorithms and scaling laws. IEEE Control Syst. Mag. **27**(4), 57–74 (2007)
4. Barooah, P., Hespanha, J.P.: Estimation from relative measurements: electrical analogy and large graphs. IEEE Trans. Signal Process. **56**(6), 2181–2193 (2008)
5. Barooah, P., Hespanha, J.P.: Error scaling laws for linear optimal estimation from relative measurements. IEEE Trans. Inf. Theory **55**(12), 5661–5673 (2009)
6. Bolognani, S., Del Favero, S., Schenato, L., Varagnolo, D.: Consensus-based distributed sensor calibration and least-square parameter identification in WSNs. Int. J. Robust Nonlinear Control **20**(2), 176–193 (2010)
7. Carron, A., Todescato, M., Carli, R., Schenato, L.: An asynchronous consensus-based algorithm for estimation from noisy relative measurements. IEEE Trans. Control Network Syst. **1**(3), 283–295 (2014)
8. Chandra, A.K., Raghavan, P., Ruzzo, W.L., Smolensky, R., Tiwari, P.: The electrical resistance of a graph captures its commute and cover times. Comput. Complex. **6**(4), 312–340 (1996)
9. Clark, A., Bushnell, L., Poovendran, R.: A supermodular optimization framework for leader selection under link noise in linear multi-agent systems. IEEE Trans. Autom. Control **59**(2), 283–296 (2014)
10. Como, G., Fagnani, F.: From local averaging to emergent global behaviors: the fundamental role of network interconnections. Syst. Control Lett. **95**, 70–76 (2016)
11. Doyle, P.G., Snell, J.L.: Random Walks and Electric Networks. Carus Monographs, Mathematical Association of America (1984)
12. Fitch, K., Leonard, N.E.: Joint centrality distinguishes optimal leaders in noisy networks. IEEE Trans. Control Network Syst. **3**(4), 366–378 (2016)
13. Frasca, P., Ishii, H., Ravazzi, C., Tempo, R.: Distributed randomized algorithms for opinion formation, centrality computation and power systems estimation. Eur. J. Control **24**, 2–13 (2015)
14. Frasca, P., Ravazzi, C., Tempo, R., Ishii, H.: Gossips and prejudices: ergodic randomized dynamics in social networks. In: IFAC Workshop on Estimation and Control of Networked Systems, Koblenz, Germany, pp. 212–219, September 2013
15. Freris, N.M., Zouzias, A.: Fast distributed smoothing of relative measurements. In: IEEE Conference on Decision and Control, Maui, HI, USA, pp. 1411–1416, December 2012
16. Friedkin, N.E., Johnsen, E.C.: Social influence networks and opinion change. In: Lawler, E.J., Macy, M.W. (eds.) Advances in Group Processes, vol. 16, pp. 1–29. JAI Press (1999)

17. Ghosh, A., Boyd, S., Saberi, A.: Minimizing effective resistance of a graph. SIAM Rev. **50**(1), 37–66 (2008)
18. Giridhar, A., Kumar, P.R.: Distributed clock synchronization over wireless networks: algorithms and analysis. In: IEEE Conference on Decision and Control, San Diego, CA, USA, pp. 4915–4920, December 2006
19. Ji, M., Ferrari-Trecate, G., Egerstedt, M., Buffa, A.: Containment control in mobile networks. IEEE Trans. Autom. Control **53**(8), 1972–1975 (2008)
20. Levin, D.A., Peres, Y., Wilmer, E.L.: Markov Chains and Mixing Times. American Mathematical Society (2009)
21. Lin, F., Fardad, M., Jovanovic, M.R.: Algorithms for leader selection in stochastically forced consensus networks. IEEE Trans. Autom. Control **59**(7), 1789–1802 (2014)
22. Lovisari, E., Garin, F., Zampieri, S.: Resistance-based performance analysis of the consensus algorithm over geometric graphs. SIAM J. Control Optim. **51**(5), 3918–3945 (2013)
23. Mobilia, M.: Does a single zealot affect an infinite group of voters? Phys. Rev. Lett. **91**(2), 028701 (2003)
24. Osting, B., Brune, C., Osher, S.J.: Optimal data collection for improved rankings expose well-connected graphs. J. Mach. Learn. Res. **15**, 2981–3012 (2014)
25. Parsegov, S.E., Proskurnikov, A.V., Tempo, R., Friedkin, N.E.: Novel multidimensional models of opinion dynamics in social networks. IEEE Trans. Autom. Control **62**(5), 2270–2285 (2017)
26. Proskurnikov, A.V., Tempo, R.: A tutorial on modeling and analysis of dynamic social networks. Part I. In: Annual Reviews in Control, pp. 65–79, March 2017
27. Ravazzi, C., Frasca, P., Ishii, H., Tempo, R.: A distributed randomized algorithm for relative localization in sensor networks. In: European Control Conference, Zürich, Switzerland, pp. 1776–1781, July 2013
28. Ravazzi, C., Frasca, P., Tempo, R., Ishii, H.: Ergodic randomized algorithms and dynamics over networks. IEEE Trans. Control Network Syst. **2**(1), 78–87 (2015)
29. Rossi, W.S., Frasca, P.: The harmonic influence in social networks and its distributed computation by message passing (2016). arXiv:1611.02955
30. Rossi, W.S., Frasca, P., Fagnani, F.: Transient and limit performance of distributed relative localization. In: IEEE Conference on Decision and Control, Maui, HI, USA, pp. 2744–2748, December 2012
31. Rossi, W.S., Frasca, P., Fagnani, F.: Average resistance of toroidal graphs. SIAM J. Control Optim. **53**(4), 2541–2557 (2015)
32. Rossi, W.S., Frasca, P., Fagnani, F.: Distributed estimation from relative and absolute measurements. IEEE Trans. Autom. Control, **PP**(99), 1, (2017). https://doi.org/10.1109/TAC.2017.2661400
33. Siami, M., Motee, N.: Fundamental limits and tradeoffs on disturbance propagation in linear dynamical networks. IEEE Trans. Autom. Control **61**(12), 4055–4062 (2016)
34. Vassio, L., Fagnani, F., Frasca, P., Ozdaglar, A.: Message passing optimization of harmonic influence centrality. IEEE Trans. Control Network Syst. **1**(1), 109–120 (2014)
35. Yildiz, E., Ozdaglar, A., Acemoglu, D., Saberi, A., Scaglione, A.: Binary opinion dynamics with stubborn agents. ACM Trans. Econ. Comput. **1**(4), 1–30 (2013)

Index

© Springer International Publishing AG 2018
F. Fagnani and P. Frasca, *Introduction to Averaging Dynamics over Networks*, Lecture Notes in Control and Information Sciences 472,
https://doi.org/10.1007/978-3-319-68022-4

Printed by Printforce, the Netherlands